Ship S
Notes &

To my wife Hilary and our family

Ship Stability Notes & Examples
Third Edition

Kemp & Young

Revised by

Dr. C. B. Barrass

OXFORD AUCKLAND BOSTON JOHANNESBURG MELBOURNE NEW DELHI

Butterworth-Heinemann
Linacre House, Jordan Hill, Oxford OX2 8DP
225 Wildwood Avenue, Woburn, MA 01801-2041
A division of Reed Educational and Professional Publishing Ltd

A member of the Reed Elsevier plc group

First published by Stanford Maritime Ltd 1959
Second edition (metric) 1971
Reprinted 1972, 1974, 1977, 1979, 1982, 1984, 1987
First published by Butterworth-Heinemann 1989
Reprinted 1990, 1995, 1996, 1997, 1998, 1999
Third edition 2001

Transferred to digital printing 2004

© P. Young 1971
C. B. Barrass 2001

All rights reserved. No part of this publication may be
reproduced in any material form (including photocopying
or storing in any medium by electronic means and whether
or not transiently or incidentally to some other use of this
publication) without the written permission of the copyright
holder except in accordance with the provisions of the Copyright,
Designs and Patents Act 1988 or under the terms of a licence
issued by the Copyright Licensing Agency Ltd, 90 Tottenham
Court Road, London, W1P 9HE, England. Applications for the
copyright holder's written permission to reproduce any part of
this publication should be addressed to the publishers

British Library Cataloguing in Publication Data
A catalogue record for this book is available from the British Library

Library of Congress Cataloguing in Publication Data
A catalogue record for this book is available from the Library of Congress

ISBN 0 7506 4850 3
Typeset by Laser Words, Madras, India

FOR EVERY TITLE THAT WE PUBLISH, BUTTERWORTH-HEINEMANN
WILL PAY FOR BTCV TO PLANT AND CARE FOR A TREE.

Contents

Preface		ix
Useful formulae		xi
Ship types and general characteristics		xv
Ship stability – the concept		xvii

I	**First Principles**		1
	Length, mass, force, weight, moment etc.		
	Density and buoyancy		
	Centre of Buoyancy and Centre of Gravity		
	Design co-effts : C_b, C_m, C_w, C_p and C_D		
	TPC and fresh water allowances		
	Permeability 'μ' for tanks and compartments		
	Fulcrums and weightless beams		
II	**Simpson's Rules – Quadrature**		19
	Calculating areas using 1st, 2nd and 3rd rules		
	VCGs and LCGs of curved figures		
	Simpsonising areas for volumes and centroids		
	Comparison with Morrish's rule		
	Sub-divided common intervals		
	Moment of Inertia about amidships and LCF		
	Moments of Inertia about the centreline		
III	**Bending of Beams and Ships**		36
	Shear force and bending moment diagrams for beams		
	Strength diagrams for ships		
IV	**Transverse Stability (Part 1)**		51
	KB, BM, KM, KG and GM concept of ship stability		
	Proof of $BM = I/V$		
	Metacentric diagrams		
	Small angle stability – angles of heel up to 15°		
	Large angle stability – angles of heel up to 90°		
	Wall-sided format for GZ		
	Stable, Unstable and Neutral Equilibrium		
	Moment of weight tables		
	Transverse Stability (Part 2)		74
	Suspended weights		

Inclining experiment/stability test
Deadweight–moment curve – diagram and use of
Natural rolling period T_R – 'Stiff' and 'tender' ships
Loss of ukc when static vessel heels
Loss of ukc due to Ship Squat
Angle of heel whilst a ship turns

V   Longitudinal Stability, i.e. Trim   89
TPC and MCT 1 cm
Mean bodily sinkage, Change of Trim and Trim ratio
Estimating new end drafts
True mean draft
Bilging an end compartment
Effect on end drafts caused by change of density

VI   Dry-docking Procedures   107
Practical considerations of docking a ship
Upthrust 'P' and righting moments
Loss in GM

VII   Water and Oil Pressure   111
Centre of Gravity and Centre of Pressure
Thrust and resultant thrust on lockgates and bulkheads
Simpson's rules for calculating centre of pressure

VIII   Free Surface Effects   119
Loss in GM, or Rise in G effects
Effect of transverse subdivisions
Effect of longitudinal subdivisions

IX   Stability Data   125
Load line rules for minimum GM and minimum GZ
Areas enclosed within a statical stability (S/S) curve
Seven parts on an statical stability (S/S) curve
Effects of greater freeboard and greater beam on an S/S curve
Angle of Loll and Angle of List comparisons
KN cross curves of stability
Dynamical stability and moment of statical stability

X   Carriage of Stability Information   138
Information supplied to ships
Typical page from a ship's Trim & Stability book
Hydrostatic Curves – diagram and use of
Concluding remarks

Contents

Appendix I	Revision one-liners	147
Appendix II	Problems	150
Appendix III	Answers to the 50 problems in Appendix II	158
Appendix IV	How to pass exams in Maritime Studies	161
Index		163

Preface

Captain Peter Young and Captain John Kemp wrote the first edition of this book way back in 1959. It was published by Stanford Marine Ltd. After a second edition (metric) in 1971, there were seven reprints, from 1972 to 1987. It was then reprinted in 1989 by Butterworth-Heinemann. A further five reprints were then made in the 1990s. It has been decided to update and revise this very popular textbook for the new millennium. I have been requested to undertake this task. Major revision has been made.

This book will be particularly helpful to Masters, Mates and Engineering Officers preparing for their SQA/MSA exams. It will also be of great assistance to students of Naval Architecture/Ship Technology on ONC, HNC, HND courses, and initial years on undergraduate degree courses. It will also be very good as a quick reference aid to seagoing personnel and shore-based staff associated with ship handling operation.

The main aim of this book is to help students pass exams in Ship Stability by presenting 66 worked examples together with another 50 exercise examples with final answers only. With this book 'practice makes perfect'. Working through this book will give increased understanding of this subject.

All of the worked examples show the quickest and most efficient method to a particular solution. Remember, in an exam, that time and inaccuracy can cost marks. To assist students I have added a section on, 'How to pass exams in Maritime Studies'. Another addition is a list of 'Revision one-liners' to be used just prior to sitting the exam.

For overall interest, I have added a section on Ship types and their Characteristics to help students to appreciate the size and speed of ships during their career in the shipping industry. It will give an awareness of just how big and how fast these modern ships are.

In the past editions comment has been made regarding Design coefficients, GM values, Rolling periods and Permeability values. In this edition, I have given typical up-to-date merchant ship values for these. To give extra assistance to the student, the useful formulae page has been increased to four pages.

Ten per cent of the second edition has been deleted. This was because several pages dealt with topics that are now old-fashioned and out of date. They have been replaced by Ship squat, Deadweight-Moment diagram, Angle of heel whilst turning, and Moments of Inertia via Simpson's Rules. These are more in line with present day exam papers.

Finally, it only remains for me to wish you the student, every success with your Maritime studies and best wishes in your chosen career.

C. B. Barrass

Useful Formulae

$KM = KB + BM$, $KM = KG + GM$

$R.D. = \dfrac{\text{Density of the substance}}{\text{Density of fresh water}}$, $\text{Displacement} = V \times \rho$

$C_b = \dfrac{V}{L \times B \times d}$, $C_w = \dfrac{WPA}{L \times B}$, $C_m = \dfrac{\text{\ss{} area}}{B \times d}$

$C_p = \dfrac{V}{\text{\ss{} area} \times L}$, $C_p = \dfrac{C_B}{C_m}$, $C_D = \dfrac{dwt}{\text{displacement}}$

$TPC = \dfrac{WPA}{100} \times \rho$, $TPC_{FW} = \dfrac{WPA}{100}$, $TPC_{SW} = \dfrac{WPA}{97.57}$

$FWA = \dfrac{\text{\textcircled{S}}}{4 \times TPC}$, $\dfrac{\text{Sinkage due to bilging}}{} = \dfrac{\text{vol. of lost buoyancy}}{\text{intact WPA}}$

permeability 'μ' $= \dfrac{BS}{SF} \times \dfrac{100}{1}$

SIMPSON'S RULES FOR AREAS, VOLUMES, C.G.S ETC.

1st rule: Area $= \tfrac{1}{3}(1, 4, 2, 4, 1)$ etc. $\times$ h

2nd rule: Area $= \tfrac{3}{8}(1, 3, 3, 2, 3, 3, 1)$ etc. $\times$ h

3rd rule: Area $= \tfrac{1}{12}(5, 8, -1)$ etc. $\times$ h

$LCG = \dfrac{\Sigma_2}{\Sigma_1} \times h$, $LCG = \dfrac{\Sigma_M}{\Sigma_A} \times \dfrac{h}{2}$

Morrish's formula: $KB = d - \dfrac{1}{3}\left(\dfrac{d}{2} + \dfrac{V}{A}\right)$

SIMPSON'S RULES FOR MOMENTS OF INERTIA ETC.

$I_{\text{\ss}} = \dfrac{1}{3} \times \Sigma_3 \times (CI)^3 \times 2$, $I_{LCF} = I_{\text{\ss}} - A(LCF)^2$

$I_{\text{\pounds}} = \dfrac{1}{9} \times \Sigma_1 \times CI \times 2$

$BM = \dfrac{I}{V}$; $BM_L = \dfrac{I_{LCF}}{V}$, $BM_T = \dfrac{I_{\text{\pounds}}}{V}$

$R_a + R_b$ = total downward forces

Anticlockwise moments = clockwise moments, $\dfrac{f}{y} = \dfrac{M}{I}$

BM for a *box-shaped* vessel; $BM_T = \dfrac{B^2}{12 \times d}$, $BM_L = \dfrac{L^2}{12 \times d}$

BM for a *triangular-shaped* vessel; $BM_T = \dfrac{B^2}{6 \times d}$, $BM_L = \dfrac{L^2}{6 \times d}$

For a *ship-shaped* vessel: $BM_T \simeq \dfrac{C_w^2 \times B^2}{12 \times d \times C_B}$ and $BM_L \simeq \dfrac{3 \times C_w^2 \times L^2}{40 \times d \times C_B}$

For a *box-shaped* vessel, $KB = \dfrac{d}{2}$ at each WL

For a *triangular-shaped* vessel, $KB = \tfrac{2}{3} \times d$ at each WL

For a *ship-shaped* vessel, $KB \simeq 0.535 \times d$ at each WL

$GZ = GM \times \sin\theta$. Righting moment $= W \times GZ$

$GZ = \sin\theta(GM + \tfrac{1}{2} \cdot BM \cdot \tan^2\theta)$

tan angle of loll $= \sqrt{\dfrac{-2 \times GM}{BM}}$

GG_1 (when loading) $= \dfrac{w \times d}{W + w}$

GG_1 (when discharging) $= \dfrac{w \times d}{W - w}$

GG_1 (when shifting) $= \dfrac{w \times d}{W}$

New KG $= \dfrac{\text{Sum of the moments about the keel}}{\text{Sum of the weights}}$

$GG_1 = GM \times \tan\theta$

$GM = \dfrac{w \times d}{W \times \tan\theta}$ or $\dfrac{w \times d}{W \times \dfrac{x}{l}}$

Maximum Kg $= \dfrac{\text{deadweight-moment}}{\text{deadweight}}$

$T_R = 2\pi\sqrt{\dfrac{k^2}{GM.g}}$, $T_R \simeq 2\sqrt{\dfrac{k^2}{GM}}$

$k \simeq 0.35 \times Br.Mld$

Loss of ukc due to heeling of *static* ship $= \tfrac{1}{2} \times b \times \sin\theta$

Useful formulae

Maximum squat 'δ' $= \dfrac{C_B \times V_K^2}{100}$ in open water conditions

Maximum squat 'δ' $= \dfrac{C_B \times V_K^2}{50}$ in confined channels

$F = 580 \times A_R \times v^2$, $\quad F_t = F \cdot \sin\alpha \cdot \cos\alpha$

$\tan\theta = \dfrac{F_t \times NL}{W \times g \times GM}$, $\quad \tan\theta = \dfrac{v^2 \times BG}{g \times r \times GM}$

MCT 1 cm $= \dfrac{W \times GM_L}{100 \times L}$, $\quad GM_L \simeq BM_L$

For an Oil Tanker, $\quad$ MCT 1 cm $\simeq \dfrac{7.8 \times (TPC)^2}{B}$

Mean bodily sinkage $= \dfrac{w}{TPC}$

Change of trim $= \dfrac{\sum(w \times d)}{MCT\ 1\ cm}$

Final Draft = Original draft + Mean bodily sinkage/rise $\pm$ for'd or aft trim ratio

$P = \dfrac{Trim \times MCT\ 1\ cm}{a}$, for drydocking

$P = \dfrac{Amount\ water\ has\ fallen\ in\ cms}{} \times TPC$, for drydocking

$\dfrac{Loss\ in\ GM\ due}{to\ drydocking} = \dfrac{P \times KM}{W}$ or $\dfrac{P \times KG}{W - P}$

Righting Moment = Effective GM $\times$ P (upthrust)

Thrust = $A \times h \times \rho$

Centre of Pressure on a bulkhead $= \tfrac{2}{3} \times h$ for *rectangular* bhd

$\qquad\qquad\qquad\quad = \tfrac{1}{2} \times h$ for *triangular* bhd

$\qquad\qquad\qquad\quad = \dfrac{\sum_2^3}{\sum_2} \times CI$ for a ship-shape bhd

Loss of GM due to FSE $= \dfrac{i \times \rho_t}{W \times n^2}$

Change in GZ $= \pm GG_1 \cdot \sin\theta$, $\quad$ 1 radian = 57.3°

Using KN Cross Curves; $\quad GZ = KN - KG \cdot \sin\theta$

Moment of statical stability = W × GZ

Dynamical stability = area under statical stability curve × W

$$= \frac{1}{3} \times \sum_1 \times CI \times W$$

Change of Trim = $\dfrac{W \times (LCG_{\cancel{R}} - LCB_{\cancel{R}})}{MCT\ 1\ cm}$

Ship Types and General Characteristics

The table below indicates characteristics relating to several types of Merchant Ships in operation at the present time. The first indicator for size of a ship is usually the Deadweight (DWT). This is the weight a ship actually carries. With some designs, like Passenger Liners, it can be the Gross Tonnage (GT). With Gas Carriers it is usually the maximum cu.m. of gas carried. Other indicators for the size of a vessel are the LBP and the block coefficient (C_b).

Type of ship or name	Typical DWT tonnes or cu.m.	LBP m	Br. Mld. m	Typical C_b fully loaded	Service speed knots
Medium sized oil tankers	50 000 to 100 000	175 to 250	25 to 40	0.800 to 0.820	15 to 15.75
ULCC, VLCC, Supertankers	100 000 to 565 000	250 to 440	40 to 70	0.820 to 0.850	13 to 15.75
OBO carriers,	up to 173 000	200 to 300	up to 45	0.780 to 0.800	15 to 16
Ore carriers. (see overpage)	up to 322 000	200 to 320	up to 58	0.790 to 0.830	14.5 to 15.5
General cargo ships	3 000 to 15 000	100 to 150	15 to 25	0.675 to 0.725	14 to 16
LNG and LPG ships	75 000 to 138 000 m³	up to 280	25 to 46	0.660 to 0.680	16 to 20.75
Container ships (see overpage)	10 000 to 72 000	200 to 300	30 to 45	0.560 to 0.600	20 to 28
Roll-on/Roll-off car and passenger ferries	2000 to 5000	100 to 180	21 to 28	0.550 to 0.570	18 to 24
Passenger Liners (see overpage)	5000 to 20 000	200 to 311	20 to 48	0.600 to 0.640	22 to 30
'QEII' (built 1970)	15 520	270	32	0.600	28.5
Oriana (built 1994)	7270	224	32.2	0.625	24
Stena Explorer SWATH (built 1996) (see overpage)	1500	107.5	40	not applicable	40

Generally, with each ship-type, an increase in the specified Service Speed for a new ship will mean a decrease in the block coefficient C_b at Draft Moulded. Since around 1975, with Very

Large Crude Carriers (VLCCs) and with Ultra Large Crude Carriers (ULCCs), there has been a gradual reduction in their designed L/B ratios. This has changed from a range of 6.0 to 6.3 to some being as low as 5.0.

One such vessel is the *Diamond Jasmine* (built in 1999), a 281 000 tonne-dwt VLCC. Her L/B is 319 m/60 m giving a ratio of 5.32. Another example commercially operating in this year 2000 is the *Chevron South America*. She is 413 160 tonnes dwt, with an LBP of 350 m and a Breadth Mld of 70 m.

One reason for these short tubby tankers is that because of safety/pollution concerns, they now have to have a double-skin hull with side ballast tanks. This of course means that for new ship orders there is an increase in breadth moulded.

To give the reader some idea of the tremendous size of ships that have been actually built the following merchant ships have been selected. Ships after all are the largest moving structures designed and built by man.

Biggest oil tanker – Jahre Viking built in 1980   Dwt = 564 739 tonnes

Seawise Giant (1980), renamed *Happy Giant* (1989), renamed *Jahre Viking* (1990)

LBP = 440 m which is approximately the length of five football or six hockey pitches!!

Br Mld = 68.80 m	SLWL = 24.61 m	Service Speed = 13 kts
Biggest ore carrier LBP = 320 m Service Speed = 14.70 kts @ 85% MCR	*Peene Ore* built in 1997 Br.Mld = 58 m	Dwt = 322 398 tonnes SLWL = 23 m
Biggest container ship LBP = 283.8 m Service Speed = 23 kts	*Nyk Antares* built in 1997 Br.Mld = 40 m	Dwt = 72 097 tonnes SLWL = 13 m
Biggest passenger liner Gross tonnage 142 000 LBP = 311 m Service Speed = 22 kts No. of Passengers = 3 114	*Voyager of the Seas* built in 1999. Br.Mld = 48 m	SLWL = 8.84 m No. of Crew = 1180 No. of Decks = 15
Fast passenger ship Ship with a twin hull (SWATH design) LBP = 107.5 m Service Speed = 40 kts	*Stena Explorer* Built in 1996 Br.Mld = 40 m Depth to Main Deck = 12.5 m	Dwt = 1500 tonnes LOA = 125 m SLWL = 4.5 m No. of Passengers = 1500

Ship Stability – the Concept

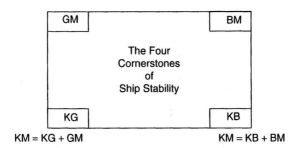

KB and BM depend on *Geometrical form* of ship.

KG depends on *loading* of ship.

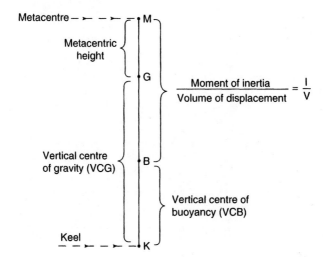

CHAPTER ONE

First Principles

SI units bear many resemblances to ordinary metric units and a reader familiar with the latter will have no difficulty with the former. For the reader to whom both are unfamiliar the principal units which can occur in this subject are detailed below.

Length

$$1 \text{ metre (m)} = 10 \text{ decimetres (dm)} = 1 \text{ m}$$
$$1 \text{ decimetre (dm)} = 10 \text{ centimetres (cm)} = 0.1 \text{ m}$$
$$1 \text{ centimetre (cm)} = 10 \text{ millimetres (mm)} = 0.01 \text{ m}$$
$$1 \text{ millimetre (mm)} \qquad\qquad\qquad = 0.001 \text{ m}$$

Mass

$$1000 \text{ grammes (g)} = 1 \text{ kilogramme (kg)}$$
$$1000 \text{ kilogrammes (kg)} = 1 \text{ metric ton}$$

Force

Force is the product of mass and acceleration

units of mass = kilogrammes (kg)

units of acceleration = metres per second squared (m/s^2)

units of force = $kg\,m/s^2$ or Newton (N) when acceleration is $9.81\,m/s^2$ (i.e. due to gravity).

$kg\,9.81\,m/s^2$ may be written as kgf so $1\,kgf = 9.81\,N$.

Weight is a force and is the product of mass and acceleration due to the earth's gravity and strictly speaking should be expressed in Newtons (N) or in mass-force units (kgf) however, through common usage the force (f) portion of the unit is usually dropped so that weight is expressed in the same units as mass.

$$1000 \text{ kgf} = 1 \text{ metric ton force} = 1 \text{ tonne}$$

So that 1 tonne is a measure of 1 metric ton weight.

Moment is the product of force and distance.

$$\text{units of moment} = \text{Newton-metre (Nm)}$$

as 9.81 Newton = 1 kgf

9810 Newton = 1000 kgf = 1 tonne.

Therefore moments of the larger weights may be conveniently expressed as tonnes metres.

Pressure is thrust or force per unit area and is expressed as kilogramme-force units per square metre or per square centimetre (kgf/m^2 or kgf/cm^2). The larger pressures may be expressed as tonnes per square metre (t/m^2).

Density is mass per unit volume usually expressed as kilogrammes per cubic metre (kg/m^3) or grammes per cubic centimetre (g/cm^3). The density of fresh water is $1\,g/cm^3$ or $1000\,kg/m^3$.

Now 1 metric ton = 1000 kg = 1 000 000 g

which will occupy   1 000 000 cm^3

but 1 000 000 cm^3 = 100 cm × 100 cm × 100 cm

= 1 m^3

so 1 metric ton of fresh water occupies 1 cubic metre. Thus numerically, $t/m^3 = g/cm^3$. Density of a liquid is measured with a Hydrometer. Three samples are usually tested and an average reading is used.

Relative density was formerly, and is still sometimes referred to as, specific gravity. It is the density of the substance compared with the density of fresh water.

$$R.D. = \frac{\text{Density of the substance}}{\text{Density of fresh water}}$$

As density of fresh water is unity ($1\,t/m^3$), the relative density of a substance is numerically equal to its density when SI units are used.

Archimedes stated that every floating body displaces its own weight of the liquid in which it floats.

It is also a fact that when a body is placed in a liquid the immersed portion of the body will displace its own volume of the liquid. If the body displaces its own weight of the liquid before it displaces its own total volume then it will float in that liquid, otherwise it will sink.

Saltwater has a relative density of 1.025 thus 1.025 metric tons of salt water occupy 1 cubic metre or 1 metric ton of salt water occupies 0.9757 cubic metres.

Iron has a relative density of 7.8 thus 7.8 tonnes of iron occupy 1 cubic metre or 1 tonne of iron occupies 0.1282 cubic metres.

If one cubic metre of iron is immersed in fresh water it will displace one cubic metre of the water which weighs 1 tonne. As 1 cubic metre of iron weighs 7.8 tonnes it is clearly not displacing its own weight. Now consider the same weight of iron with an enlarged volume, say

First principles

2 cubic metres (an air space of 1 cubic metre having been introduced in the centre of the iron). If this enlarged volume is immersed in fresh water 2 cubic metres are displaced and these 2 cubic metres weigh 2 tonnes. There is still insufficient weight of fresh water being displaced for the iron to float so the volume of the iron will have to be further increased – without increase in weight – if the iron is to float. When the volume of the iron (and air space) reaches $7.8\,m^3$, $7.8\,m^3$ of fresh water will be displaced and this weighs 7.8 tonnes which is exactly equal to the weight of the piece of iron. The iron will now just float. If the volume of the iron is increased still further it will float with a certain amount of freeboard as, if the volume were to be completely immersed a weight of fresh water more than the weight of the iron would have been displaced.

We can now summarise by saying that if the R.D. of the body taken as a whole is less than the R.D. of the liquid in which it is placed, then it will float in that liquid.

Reserve buoyancy is virtually the watertight volume above the waterline. It is necessary to have a certain reserve of buoyancy as, when in a seaway with the ends or the middle unsupported, the vessel will sink down to displace the same volume as she does when in smooth water. This could result in the vessel being overwhelmed. This is illustrated below.

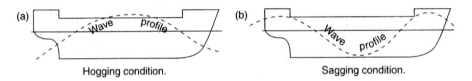

(a) Hogging condition.   (b) Sagging condition.

Figure 1.1

Centre of Buoyancy (KB) is the geometrical centre of the underwater volume and the point through which the total force due to buoyancy may be considered to act vertically upwards. Let d = draft.

In a boxshape	KB is $0.5\,d$
In a triangular shape	KB is $2/3\,d$
In a shipshape	KB is approximately $0.535 \times d$

above the keel

The position of the centre of buoyancy may be calculated by Simpson's Rules as shown in Chapter 2. The approximate position may also be found by Morrish's Formula which is:

$$\text{C of B below waterplane} = \frac{1}{3}\left(\frac{d}{2} + \frac{V}{A}\right) \quad \text{where} \begin{cases} d \text{ is the draft} \\ V \text{ is the volume of displacement} \\ A \text{ is area of the waterplane} \end{cases}$$

Centre of Gravity (G) is that point in a body through which the total weight of the body may be considered to be acting. (It will be useful to remember that the resultant moment about the C of G is zero.)

The methods of finding and calculating the position of G are given in Chapter 4.

Design co-efficients: The Naval Architect uses many co-efficients in ship technology, five of which are listed below:

1. Block co-efft (C_b)... or co-efft of Fineness
2. Waterplane Area co-efft (C_w)
3. Midship Area co-efft (C_m or $C_\times$)
4. Prismatic co-efft (C_p)
5. Deadweight co-efft (C_D)

Block co-efft (C_b) is the ratio between the underwater volume (V) and the volume of the circumscribing block.

$$C_b = \frac{V}{L \times B \times d} \quad \left\{ \begin{array}{l} \text{full-form, if } C_b > 0.700 \\ \text{medium form, if } C_b \simeq 0.700 \\ \text{fine-form, if } C_b < 0.700 \end{array} \right\}$$

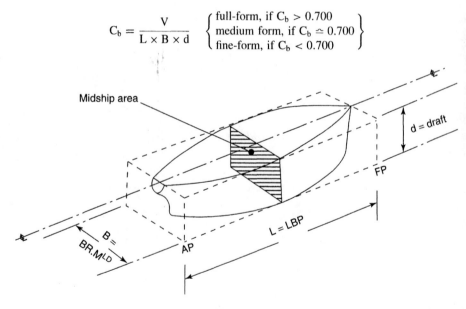

Figure 1.2

For merchant ships C_b will range, depending upon ship type, from about 0.500 up to 0.850. See later table.

Waterplane Area co-efft (C_w) is the ratio between the waterplane area (WPA) and the area of the surrounding rectangle.

$$C_w = \frac{WPA}{L \times B}.$$

First principles

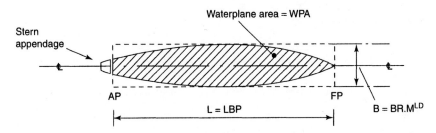

Figure 1.3

When ships are fully-loaded, a useful approximation for this waterplane area co-efft is:

$$C_w = \left(\tfrac{2}{3} \times C_b\right) + \tfrac{1}{3}. \text{ @ fully-loaded draft } only.$$

At drafts below SLWL, the WPA decreases and with it the C_w values.

Midship Area co-efft (C_m) is the ratio of the midship area and the surrounding rectangle of (B × d).

$$\therefore C_m = \frac{\text{midship area}}{B \times d}$$

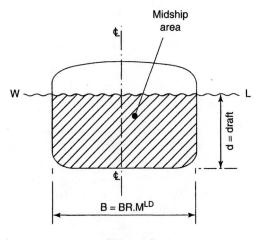

Figure 1.4

For merchant ships, C_m will be of the order of 0.975 to 0.995, when fully-loaded.

Prismatic co-efft (C_p) is the ratio of the underwater volume (V) and the multiple of midship area and the ship's length

$$\therefore C_p = \frac{V}{\text{midship area} \times L}.$$

C_p is a co-efft used mainly by researchers working with ship-models at a towing tank. If we divide C_B by C_m we obtain:

$$\frac{C_B}{C_m} = \frac{V}{L \times B \times d} \times \frac{B \times d}{\text{midship area}} = \frac{V}{\text{midship area} \times L}$$

Hence $C_p = \frac{C_B}{C_m}$ at each waterline. Consequently, C_p will be just above C_B value, for each waterline.

Deadweight co-efft (C_D) is the ratio of a ship's deadweight (carrying capacity in tonnes) with the ship's displacement (W)

$$\therefore C_D = \frac{\text{Deadweight}}{W}$$

Summary for the design co-effts: First of all remember that all these co-effts will *never be more than unity*. The table below indicates typical C_b values for several ship types.

Ship type	Typical C_b fully-loaded	Ship type	Typical C_b fully-loaded
ULCC	0.850	General Cargo ships	0.700
Supertankers	0.825	Passenger Liners	0.625
Oil tankers	0.800	Container ship/RoRo	0.575
Bulk carriers	0.750	Tugs	0.500

With Supertankers and ULCCs, it is usual to calculate these design co-effts to four decimal figures. For all other ship types, sufficient accuracy is obtained by rounding off to three decimal figures.

The table below indicates typical C_D values for several ship types.

Ship type	Typical C_D fully-loaded	Ship type	Typical C_D fully-loaded
Oil Tankers	0.800 to 0.860	Container ships	0.600
Ore-carriers	0.820	Passenger Liners	0.350 to 0.400
General Cargo	0.700	RoRo vessels	0.300
LNG/LPG carriers	0.620	Cross-Channel ferries	0.200

When fully-loaded, for Oil Tankers and General Cargo ships C_D and C_B will be very close, the former being slightly the higher in value.

WORKED EXAMPLE 1

An Oil Tanker has a Breadth Moulded of 39.5 m with a Draft Moulded of 12.75 m and a midship area of 496 m^2.

(a) Calculate her midship area co-efft (C_m).

$$C_m = \frac{\text{⋈ area}}{B \times d} = \frac{496}{39.5 \times 12.75} = 0.9849$$

$$\therefore C_m = 0.9849.$$

(b) Calculate the bilge radius Port & Starboard.

$$\text{midship area} = \{B \times d\} - \left\{\frac{0.2146 \times r^2}{2}\right\}$$

where r = bilge radius P & S

$$\therefore \left\{\frac{\text{midship area} - (B \times d)}{0.2146}\right\} \cdot 2 = -r^2$$

$$\therefore \left\{\frac{496 - (39.5 \times 12.75)}{0.2146}\right\} = -r^2$$

$$\therefore -17.77 = -r^2$$

$$\therefore r^2 = 17.77$$

$$\therefore r = \sqrt{17.77} = 4.215 \, \text{m}.$$

WORKED EXAMPLE 2

For a General cargo ship LBP = 120 m, Breadth Mld = 20 m, draft = 8 m, displacement @ 8 m draft = 14 000 t, $C_m = 0.985$, $C_w = 0.808$. Using ship surgery, a midship portion 10 m long is welded into the ship. Calculate the new C_b, C_w, C_p and displacement values.

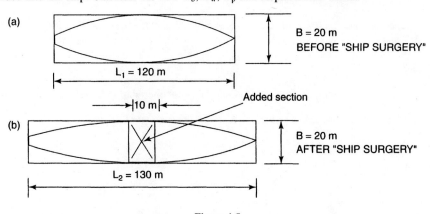

Figure 1.5

Volume of added portion = $C_m \times B \times d \times l = 0.985 \times 20 \times 8 \times 10$

$$= 1576 \, \text{m}^3$$

$$\delta W = \delta V \times \rho_{sw}$$
$$\therefore \delta W = 1576 \times 1.025 = 1620 \text{ tonnes}$$

New displacement $= W_1 + \delta W = 14\,000 + 1620 = \underline{15\,620\,tonnes} = W_2$

$$\text{New } C_b = \frac{\text{Vol of displacement}_{(2)}}{L_2 \times B \times d} = \frac{15\,620 \times 1/1.025}{130 \times 20 \times 8}$$

New $C_b = 0.733.$

New WPA $= (0.808 \times L_1 \times B) + (l \times B)$

$\quad\quad\quad\quad = (0.808 \times 120 \times 20) + (10 \times 20)$

$\therefore$ WPA$_2 = 1939.2 + 200 = 2139.2\,\text{m}^2$

$$\text{New } C_W = \frac{\text{WPA}_2}{L_2 \times B} = \frac{2139.2}{130 \times 20} = \underline{0.823}$$

$$\text{New } C_p = \frac{\text{Vol of displacement}_{(2)}}{L_2 \times \text{× area}} = \frac{15\,620 \times 1/1.025}{130 \times 0.985 \times 20 \times 8} = \underline{0.744}.$$

Check: $C_{p(2)}$ also $= \dfrac{C_{B(2)}}{C_m} = \dfrac{0.733}{0.985} = 0.744$, as above.

Approximation: $C_{W(2)} = \frac{2}{3} \cdot C_{B(2)} + \frac{1}{3} = (\frac{2}{3} \times 0.733) + \frac{1}{3} = 0.822$ i.e close to above new C_w answer.

WORKED EXAMPLE 3

A container-ship has the following C_w values commencing at the base: 0.427, 0.504, 0.577, 0.647 and 0.715 at the Summer Load Water line (SLWL). These C_w values are spaced equidistant apart up to the Draft moulded. A knowledge of Simpson's Rules is required for this example. See Chapter 2.

(a) *Calculate* the block co-efficient C_b when this container-ship is loaded up to her SLWL.

(b) *Estimate* her C_b when fully-loaded, using an approximate formula.

(a) Each WPA $= C_w \times L \times B\,\text{m}^2$ and CI $= d/4$.

	WPA (m^2)	SM	Volume function
Base	0.427·LB	1	0.427·LB
	0.504·LB	4	2.016·LB
	0.577·LB	2	1.154·LB
	0.647·LB	4	2.588·LB
SLWL	0.715·LB	1	0.715·LB
			6.900 · LB $= \sum_1$

First principles

$\sum_I$ = 'Summation of'.

$$\text{Volume of displacement} = \frac{1}{3} \times CI \times \sum_I \text{ and } C_b = \frac{\text{Vol. of displacement}}{L \times B \times d}$$

$$\therefore C_b = \frac{\frac{1}{3} \times CI \times \sum_I}{L \times B \times d} = \frac{\frac{1}{3} \times \frac{d}{4} \times 6.900 \cdot \cancel{L} \cancel{B}}{\cancel{L} \times \cancel{B} \times \cancel{d}} = \frac{6.900}{12}$$

$$\therefore C_b = 0.575, \text{ when fully-loaded.}$$

(b) *Approx formula:* $C_w = (\frac{2}{3} \times C_b) + \frac{1}{3}$

$$\therefore (C_w - \frac{1}{3}) \times \frac{3}{2} = C_b \quad @ \text{ SLWL i.e. Draft Mld.}$$

$$\therefore (0.715 - 0.333) \times \frac{3}{2} = C_b$$

$$\therefore C_b = 0.382 \times \frac{3}{2}.$$

$\therefore$ Approximately $C_b = 0.573$, which is very near to previous answer.

DISPLACEMENT

Volume of displacement in cubic metres = $L \times B \times d \times C_b$

where L = length in metres

B = breadth in metres

d = draft in metres

C_b = block coefficient (coefficient of fineness)

Displacement in tonnes = volume displaced × density

where above density is expressed in tonnes/m^3

WORKED EXAMPLE 4

A vessel of triangular form length 100 m, beam 12 m, depth 6 m is displacing 3030 tonnes in water relative density 1.010. What is her reserve buoyancy?

$$\text{Volume of Displacement} = \frac{3030}{1.010} = 3000 \, m^3$$

$$\text{Total volume of vessel} = \frac{100 \times 12 \times 6}{2} = 3600 \, m^3$$

$$\text{Reserve Buoyancy} = \text{Total volume} - \text{volume of displacement}$$

$$= 3600 - 3000$$

$$= 600 \, m^3.$$

Tonnes per centimetre immersion (TPC) is the additional tonnage displaced when the mean draft is increased by one centimetre from stern to bow.

Additional volume displaced when the draft is increased by 1 centimetre is WPA $\times \frac{1}{100}$ cubic metres where WPA is in square metres (m²).

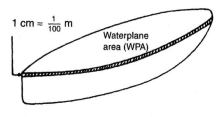

but 1 cubic metre = 1 metric ton of FW

therefore $TPC_{FW} = \dfrac{WPA}{100}$

or $TPC_{SW} = \dfrac{1.025 WPA}{100} = \dfrac{WPA}{97.57}$.

Figure 1.6

TO FIND THE FRESH WATER ALLOWANCE (FWA)

L, B, d in metres   A is waterplane area in m²   T is TPC

△F̄ FW displacement at summer draft

△S̄ SW displacement at summer draft

Then L × B × d × C_b × density = displacement in metric tons

$$△F̄ \times 1.025 = △F̄ + A \times FWA \text{ (in metres)}$$

$$△F̄ \times 0.025 = A \times FWA \text{ metres}$$

$$A \times FWA \text{ metres} = \frac{△F̄}{40}$$

$$\frac{100T}{1.025} \times FWA \text{ metres} = \frac{△F̄}{40}$$

$$FWA \text{ metres} = \frac{1.025 △F̄}{4000T}$$

$$FWA \text{ millimetres} = \frac{△S̄ \times 1000}{4000T} = \frac{△S̄}{4T}$$

N.B. The FWA is given in this form in the Load Line Rules. See Worked Example 6.

First principles

WORKED EXAMPLE 5

A vessel loads to her summer loadline at an up river port where the relative density of the water is 1.002. She then proceeds down river to a port at the river mouth where the water has relative density of 1.017, consuming 25 tonnes of fuel and water on passage. On loading a further 100 tonnes of cargo, it is noted that she is again at her summer loadline. What is her summer displacement in salt water?

$$\text{Displacement tonnes} = V \times \text{density}$$

as she is always at the same draft V is constant.

so V 1.002 = displacement up river in tonnes

V 1.017 = displacement down river in tonnes

and V 1.017 = V 1.002 + 75

$$V = \frac{75}{0.015}$$

$$V = 5000 \, m^3$$

so in salt water, Displacement = $V \times \rho_{SW}$ = 5000 × 1.025 = 5125 tonnes.

WORKED EXAMPLE 6

L = 130, Br.Mld = 19.5 m, SLWL = 8.15 m, C_m = 0.988. Fully-loaded displacement = 14 500 tonnes.

Calculate: C_b, midship area, C_p by two methods, approx C_w @ SLWL, WPA @ SLWL, TPC_{sw} @ SLWL, Fresh Water Allowance (FWA).

$$C_b = \frac{\text{Vol of displacement}}{L \times B \times d} = \frac{14\,500 \times \frac{1}{1.025}}{130 \times 19.5 \times 8.15} = 0.685.$$

midship area = $C_m \times B \times d$ = 0.988 × 19.5 × 8.15 = 157.02 m².

$$C_p = \frac{\text{Vol of displacement}}{L \times \text{area}} = \frac{14\,500 \times \frac{1}{1.025}}{130 \times 157.02} = 0.693.$$

$$\text{or } C_p = \frac{C_b}{C_m} = \frac{0.685}{0.988} = 0.693 \text{ (as above)}$$

At SLWL $C_w = \frac{2}{3}C_b + \frac{1}{3} = (\frac{2}{3} \times 0.685) + \frac{1}{3}$

∴ C_w = 0.790 @ SLWL.

$$C_w = \frac{WPA}{L \times B} \quad \therefore WPA = C_w \times L \times B$$

$$\therefore \text{WPA} = 0.790 \times 130 \times 19.5$$

$$\therefore \text{WPA} = 2003 \text{ m}^2 \text{ @ SLWL}$$

$$\text{TPC}_{\text{SW}} = \frac{\text{WPA}}{100} \times \rho_{\text{SW}} = \frac{2003}{100} \times 1.025$$

$$\text{TPC}_{\text{SW}} = 20.53 \text{ t.}$$

$$\text{FWA} = \frac{W}{4 \times \text{TPC}_{\text{SW}}} = \frac{14\,500}{4 \times 20.53} = 177 \text{ mm}$$

$$= 0.177 \text{ m.}$$

WORKED EXAMPLE 7

A 280 650 t dwt VLCC has the following characteristics:

LBP = 319 m, Br. Mld = 56 m, SLWL = 20.90 m, displacement @ SLWL = 324 522 t, $C_m = 0.9882$.

Using this data, calculate the following:

1. C_b
2. Approx C_w @ SLWL
3. WPA
4. TPC_{sw}
5. Midship area
6. C_p
7. C_D
8. Approx MCT 1 cm.

$$C_b = \frac{\text{Vol of displacement}}{L \times B \times d} = \frac{324\,522 \times \frac{1}{1.025}}{319 \times 56 \times 20.90} = 0.8480.$$

Approx C_w @ SLWL $= (\frac{2}{3} \times C_b) + \frac{1}{3} = (\frac{2}{3} \times 0.8480) + \frac{1}{3} = 0.8987.$

$$C_w = \frac{\text{WPA}}{L \times B} \quad \therefore \text{WPA} = C_w \times L \times B = 0.8987 \times 319 \times 56 = 16\,054 \text{ m}^2.$$

$$\text{TPC}_{\text{sw}} = \frac{\text{WPA}}{100} \times \rho_{\text{sw}} = \frac{16\,054}{100} \times 1.025 = 164.55 \text{ tonnes.}$$

$$C_m = \frac{\text{midship area}}{B \times d} \quad \therefore \text{midship area} = C_m \times B \times d$$

$$= 0.9882 \times 56 \times 20.90$$

$$= 1156.6 \text{ m}^2.$$

$$C_p = \frac{\text{vol of } \Delta't}{L \times \text{midship area}} = \frac{324\,522 \times \frac{1}{1.025}}{319 \times 1156.6} = 0.8581.$$

First principles

$$C_p \text{ also} = \frac{C_b}{C_m} = \frac{0.8480}{0.9882} = 0.8581 \text{ (as on previous page).}$$

$$C_D = \frac{DWT}{\Delta't} = \frac{280\,650}{324\,522} = 0.8648.$$

$$\text{approx MCT 1 cm} = \frac{7.8 \times (TPC_{sw})^2}{B} = \frac{7.8 \times 164.55^2}{56}$$

$$= 3771 \text{ t.m/cm}.$$

We have shown how a steel ship can be made to float. Suppose we now bilge the vessel's hull in way of a midship compartment, as shown below;

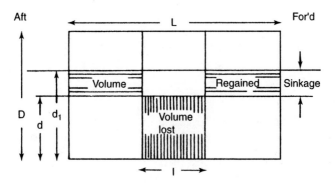

Figure 1.7

A volume of buoyancy $l \times B \times d$ is lost. The vessel will sink lower in the water until this has been replaced, and part of her reserve buoyancy will be used. The total reserve buoyancy is $(L - l) \times B \times (D - d)$, the portion which will be used is $(L - l) \times B \times (d_1 - d)$.

$$\text{i.e. } l \times B \times d = (L - l) \times B \times (d_1 - d)$$

$$\text{so } d_1 - d = \frac{l \times B \times d}{(L - l) \times B} \text{ or } \frac{l \times d}{(L - l)}$$

or

$$\frac{\text{The increase in draft due to bilging}}{} = \frac{\text{The volume of lost buoyancy}}{\text{The area of the intact waterplane}} = \frac{\text{lost buoyancy volume}}{\text{intact WPA}}$$

It can be seen that a reduction in lost buoyancy means less sinkage in the event of a compartment being bilged. The volume of lost buoyancy can be reduced either by fitting a watertight flat or by putting cargo or water ballast into the compartment. When cargo is in a compartment, only part of the volume of the compartment will be available for incoming water. The cargo will also contribute to the intact waterplane area value.

If the watertight flat is either at or below the waterline, the length of the intact waterplane will be the full length of the vessel.

Permeability 'μ' is the amount of water that can enter a bilged compartment

Empty compartment... 'μ' = 100%.   Engine Room 'μ' = 80% to 85%.
Grain filled hold...... 'μ' = 60% to 65%.   Filled water-ballast tank 'μ' = 0%.

WORKED EXAMPLE 8

A box shaped vessel length 72 m, breadth 8 m, depth 6 m, floating at a draft of 4 m has a midship compartment 12 m long. What will be the sinkage if this compartment is bilged if:

a) A watertight flat is fitted 5 m above the keel?
b) A watertight flat is fitted 4 m above the keel?
c) A watertight flat is fitted 2 m above the keel?
d) A watertight flat is fitted 4.5 m above the keel?

a) $\text{Sinkage} = \dfrac{\text{Volume of lost buoyancy}}{\text{Area of intact waterplane}}$

$= \dfrac{12 \times 8 \times 4}{(72 - 12) \times 8} = 0.8 \, \text{m}.$

Note: When using the 'Lost Buoyancy Method', W the Weight and KG remain *unchanged* after bilging has taken place.

b) $\text{Sinkage} = \dfrac{\text{Volume of lost buoyancy}}{\text{Area of the intact waterplane}}$

$= \dfrac{12 \times 8 \times 4}{72 \times 8} = 0.667 \, \text{m}.$

When using the 'Added Weight method', the Weight W and KG *do change* after bilging has taken place.

c) $\text{Sinkage} = \dfrac{\text{Volume of lost buoyancy}}{\text{Area of the intact waterplane}}$

$= \dfrac{12 \times 8 \times 2}{72 \times 8} = 0.333 \, \text{m}.$

This assumes that the hull is bilged below the flat.

d) The volume of lost buoyancy $= 12 \times 8 \times 4 = 384 \, \text{m}^3$

Intact volume between 4 m and 4.5 m $= (72 - 12) \times 8 \times 0.5$

$= 240 \, \text{m}^3.$

Volume still to be replaced $= 384 \, \text{m}^3 - 240 \, \text{m}^3 = 144 \, \text{m}^3$

$\text{Further sinkage} = \dfrac{\text{Volume still to be replaced}}{\text{Area of the waterplane (above W/T flat)}}$

$$\text{Further sinkage} = \frac{144}{72 \times 8} = 0.25 \text{ m}.$$
$$\text{Total sinkage} = 0.50 + 0.25$$
$$= 0.75 \text{ m}.$$

WORKED EXAMPLE 9

A vessel whose TPC is 12.3, is drawing 4 m. A rectangular midship cargo compartment 12 m long, 10 m breadth and 6 m depth has a permeability 'μ' of 60%. What would be the mean draft if the compartment was bilged?

$$\text{Now TPC}_{sw} = \frac{\text{WPA}}{100} \times \rho_{sw}$$

$$\therefore \text{Vessel's waterplane area} = \frac{12.3 \times 100}{1.025} = 1200 \text{ m}^2$$

$$\text{Area of compartment} = 12 \text{ m} \times 10 \text{ m} = \underline{-120 \text{ m}^2}$$

$$\text{Fully intact area} = 1080 \text{ m}^2$$

$$+(100 - \mu) \text{ of comp't area} = 40\% \times l \times b = \underline{+48 \text{ m}^2}$$

$$\text{Effective } \textit{intact} \text{ area} = \underline{1128 \text{ m}^2}$$

$$\text{Mean bodily sinkage} = \frac{\text{Volume of lost buoyancy}}{\text{Area of intact waterplane}}$$

$$= \frac{12 \times 10 \times 4 \times 60\%}{1128} = 0.26 \text{ m}.$$

$$\text{Original draft} = 4.00 \text{ m}$$
$$+ \text{ mean bodily sinkage} = +0.26 \text{ m}$$
$$\text{New draft} = \underline{4.26 \text{ m}}$$

$\therefore$ *New draft* = 4.26 m.

WORKED EXAMPLE 10

If a ship's compartment has a stowage factor of 1.50 cubic metres per tonne together with a relative density of 0.80, then estimate the permeability 'μ' if this compartment is bilged.

$$\text{Space occupied by one tonne of cargo} = \frac{1}{\text{R.D}} = \frac{1}{0.80}$$
$$= 1.25 \text{ m}^3 \ldots \ldots (\text{I})$$
$$\text{SF} = \text{Stowage factor for cargo} = 1.50 \text{ m}^3 \ldots \ldots (\text{II})$$
$$\therefore \text{Broken Stowage} = (\text{II}) - (\text{I}) = \text{BS} = \underline{0.25} \text{ m}^3.$$

Permeability 'μ' = $\dfrac{BS}{SF} \times 100$ per cent

$= \dfrac{0.25}{1.50} \times 100 = \dfrac{100}{6}$ per cent.

Thus permeability 'μ' = 16.67%.

PRINCIPLES OF TAKING MOMENTS

A moment of a force (weight) about a point can be defined as being the product of the force and the perpendicular distance of the point of application of the force from the point about which moments are being taken. It is expressed in force-distance units which for problems associated with ships will be tonnes-metres. F is the Fulcrum point, about which moments can be calculated.

A simple application of the principle of moments is that of calculating the position of the centre of gravity of a number of weights as shown below.

Consider a weightless beam AB, balanced at point F (the Fulcrum).

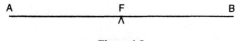

Figure 1.8

Place a weight of 5 tonnes 8 metres from F. This causes a moment of 5×8 tonnes-metres about F and consequently a rotation of the beam in a clockwise direction.

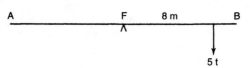

Figure 1.9

To keep the beam balanced, a similar weight of 5 tonnes may be placed at the same distance from F, but on the opposite side. The moments will now be equal and the rotational effect of the first weight counteracted.

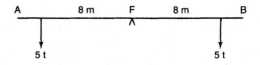

Figure 1.10

First principles

If however, there is only a weight of 8 tonnes available, we could place this at a distance of 5 metres to cause a moment which will again balance the original upsetting moment.

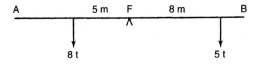

Figure 1.11

Weights of 6 tonnes and 11 tonnes at distances of 3 and 2 metres will give the same total moment on the left-hand side of Fulcrum F. There are many combinations of weight and distance which can cause a moment of 40 tonnes-metres and so keep the beam balanced.

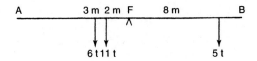

Figure 1.12

It must be clearly understood that the moment which is caused is all important, this is the product of weight and distance.

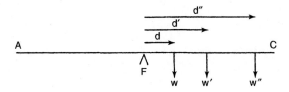

Figure 1.13

The beam AC, which is still weightless, now has weights w, w', and w" attached at distances d, d', and d" from F. To balance the beam we could put similar weights at similar distances on the opposite side of F, or we could put weight W, which is equal to the sum of w, w', w", at a distance D from F.

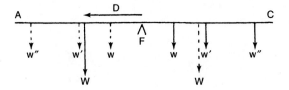

Figure 1.14

As W is replacing w, w', w", the position at which it is placed will be the centre of gravity of the weights w, w', w".

In order to find the distance D the moments each side of Fulcrum F should be equated i.e.

$$W \times D = (w \times d) + (w' \times d') + (w'' \times d'')$$

$$\text{then } D = \frac{(w \times d) + (w' \times d') + (w'' \times d'')}{W}$$

In general terms

The distance of the centre of gravity of a number of weights from the Fulcrum point about which moments are taken $= \dfrac{\text{Sum of the moments about the Fulcrum point}}{\text{Sum of the weights}}$

It should be understood that moments may be taken about any convenient 'datum' point. When taking moments aboard a ship, it is usual to take them about either the keel or the ship's centre of gravity when considering *vertical* positions of G. When considering *transverse* positions of G, moments are taken about either the centreline or the ship's centre of gravity. Examples of these are given on pages 66 and 69 in Chapter 4.

The same principles as those outlined above are used for calculating the centres of areas (2-dimensional centres) or the centres of volumes (3-dimensional centres). See pages 25 and 28 in Chapter 2. For bending moments see page 37 in Chapter 3. For trimming moments see page 90 in Chapter 5.

CHAPTER TWO

Simpson's Rules – Quadrature

An essential in many of the calculations associated with stability is a knowledge of the waterplane area at certain levels between the base and 85% × Depth Mld.

There are several ways by which this can be found, two of them being the Trapezoidal Rule and Simpson's Rules. The former rule assumes the bounding curve to be a series of straight lines and the waterplane to be divided into a number of trapezoids (a trapezoid is a quadrilateral with one pair of opposite sides parallel), whereas the latter rules assume the curves bounding the area to be parabolic. Simpson's Rules can be used to find areas of curved figures without having to use calculus.

As the waterplane is symmetrical about the centre line, it is convenient to consider only half the area. ℄ is the centreline.

Figure 2.1 shows a half waterplane area with semi-ordinates (y, y_1, y_2, etc.) so spaced that they are equidistant from one another. This distance is known as the common interval (h or CI).

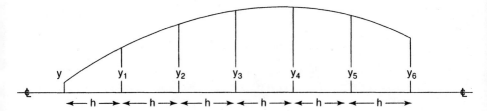

Figure 2.1 Semi-ordinates are also known as half-ordinates or offsets, in ship calculations.

By the trapezoidal rule the area = $h \times \left(\frac{y + y_6}{2} + y_1 + y_2 + y_3 + y_4 + y_5\right)$

In general terms, the area can be expressed as:

$$\left(\frac{\text{sum of the end ordinates}}{2} + \frac{\text{sum of the remaining}}{\text{ordinates}}\right) \times \text{common interval}$$

It will be noted that the accuracy increases with the number of trapezoids which are formed, that is to say the smaller the common interval the less the error. Where the shape changes rapidly (e.g. at the ends) the common interval may be halved or quartered. The trapezoidal rule is used mainly in the U.S.A. In British shipyards Simpson's Rules, of which there are three, are in common use. They can be used very easily in modern computer programmes/packages.

As examination syllabuses specifically mention Simpson's Rules the student is advised to study them carefully and to use them in preference to the trapezoidal rule.

SIMPSON'S FIRST RULE

This is to be used when the number of intervals is divisible by 2. The multipliers are 1 4 1, which become 1 4 2 4 2... 4 1 when there are more than 2 intervals. This is shown below.

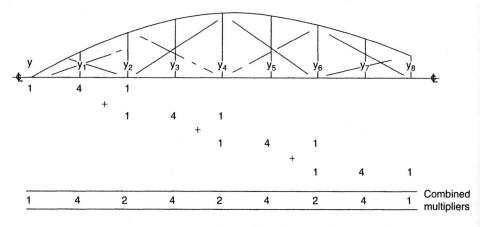

Figure 2.2

TO USE

Multiply each of the ordinates by the appropriate multiplier; this gives a product (or function) for area. Add up these products and multiply their sum by $\frac{1}{3}$ of the common interval in order to obtain the area.

$$\text{i.e. Area} = \frac{1}{3} h \times \left(y + 4y_1 + 2y_2 + 4y_3 + 2y_4 + 4y_5 + 2y_6 + 4y_7 + y_8 \right)$$

Note how the multipliers begin and end with 1.

SIMPSON'S SECOND RULE

This is to be used when the number of intervals is divisible by 3. The multipliers are 1 3 3 1, which become 1 3 3 2 3 3 2...3 3 1 when there are more than 3 intervals. This is shown on next page.

TO USE

Multiply each of the ordinates by the appropriate multiplier to give a product for area. Add up these products and multiply their sum by 3/8 of the common interval in order to obtain

Simpson's rules – quadrature

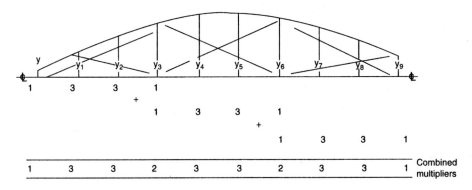

Figure 2.3 Note how the multipliers begin and end with 1

the area.

i.e. Area $= \frac{3}{8} \times h(y + 3y_1 + 3y_2 + 2y_3 + 3y_4 + 3y_5 + 2y_6 + 3y_7 + 3y_8 + y_9)$

SIMPSON'S THIRD RULE

Commonly known as the 5, 8 minus 1 rule.

This is to be used when the area between any two adjacent ordinates is required, three consecutive ordinates being given.

The multipliers are 5, 8, −1.

TO USE

To 5 times the ordinate bounding the area add 8 times the middle ordinate, subtract the other given ordinate and multiply this result by 1/12th of the common interval.

Area ABCD $= 1/12h \times (5y + 8y_1 - y_2)$

Area CDEF $= 1/12h \times (5y_2 + 8y_1 - y)$

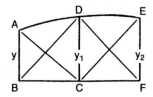

Figure 2.4

It will be noted that if we add the previous two areas together we have the first rule in a slightly different form. See Worked Example 12.

Although only areas have been mentioned up to now, Simpson's Rules are also used for calculating volumes. To do so a series of given areas is put through the rules; worked examples covering this are to be found further on.

Occasionally it will be found that either the first or the second rule can be used; in such cases it is usual to use the first rule.

When neither rule will fit the case a combination of rules will have to be used. See Worked Example 17.

WORKED EXAMPLE 11

For a Supertanker, her fully loaded waterplane has the following 1/2-ordinates spaced 45 m apart. 0, 9.0, 18.1, 23.6, 25.9, 26.2, 22.5, 15.7 and 7.2 metres respectively.

Calculate the WPA and TPC in salt water.

1/2-ord	SM	Area function
0	1	0
9.0	4	36.0
18.1	2	36.2
23.6	4	94.4
25.9	2	51.8
26.2	4	104.8
22.5	2	45.0
15.7	4	62.8
7.2	1	7.2
		$438.2 = \Sigma_1$

Σ = 'summation of'

$$\text{WPA} = \frac{1}{3} \times h \times \Sigma_1 \times 2 \text{ (for both sides)}$$

$$= \frac{1}{3} \times 45 \times 438.2 \times 2$$

$\therefore$ WPA = 13 146 m^2.

$$\text{TPC}_{sw} = \frac{\text{WPA}}{100} \times \rho_{sw}$$

$$= \frac{13\,146}{100} \times 1.025$$

$\therefore$ TPC$_{sw}$ = 134.75.

WORKED EXAMPLE 12

The 1/2-ords of a curved figure are 18.1, 23.6 and 25.9 m, spaced 24 m apart.

Calculate the area enclosed between the:

(a) 1st and 2nd 1/2-ordinates

(b) 2nd and 3rd 1/2-ordinates

(c) Check your first two answers are correct, by using Simpson's first rule.

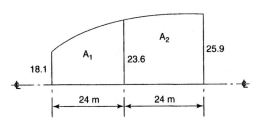

Figure 2.5

(a)

1/2-ord	SM	Area Function
18.1	5	90.5
23.6	8	188.8
25.9	−1	−25.9
		253.4 = $\sum_3$

$A_1 = 1/12 \times \sum_3 \times h \times 2$

$A_1 = 1/12 \times 253.4 \times 24 \times 2$

$\therefore A_1 = 1013.6\,m^2.$

(b)

1/2-ord	SM	Area Function
18.1	−1	−18.1
23.6	8	188.8
25.9	5	129.5
		300.2 = $\sum_3$

$A_2 = 1/12 \times \sum_3 \times h \times 2$

$A_2 = 1/12 \times 300.2 \times 24 \times 2$

$\therefore A_2 = 1200.8\,m^2.$

$\therefore A_1 + A_2 = 1013.6 + 1200.8 = 2214.4\,m^2 = A_3.$

(c)

1/2-ord	SM	Area Function
18.1	1	18.1
23.6	4	94.4
25.9	1	25.9
		138.4 = $\sum_1$

$A_3 = 1/3 \times \sum_1 \times h \times 2$

$A_3 = 1/3 \times 138.4 \times 24 \times 2$

$A_3 = 2214.4\,m^2.$

This checks out exactly with the summation of answers (a) & (b).

WORKED EXAMPLE 13

A water plane of length 270 m and breadth 35.5 m has the following equally spaced breadths: 0.3, 13.5, 27.0, 34.2, 35.5, 35.5, 35.5, 32.0, 23.1 and 7.4 m respectively.

Calculate the WPA, waterplane area co-efft and the TPC in fresh water. Use Simpson's 2nd rule.

Ordinate	SM	Area ftn
0.3	1	0.3
13.5	3	40.5
27.0	3	81.0
34.2	2	68.4
35.5	3	106.5
35.5	3	106.5
35.5	2	71.0
32.0	3	96.0
23.1	3	69.3
7.4	1	7.4
		$646.9 = \sum_2$

$\sum$ = 'summation of'

$$\text{WPA} = 3/8 \times \sum\nolimits_2 \times h$$

$$h = \frac{270}{9} = 30 \, m$$

$$\text{WPA} = 3/8 \times 646.9 \times 30 = 7278 \, m^2$$

$$C_w = \frac{\text{WPA}}{L \times B} = \frac{7278}{270 \times 35.5} = 0.759$$

$$\text{TPC in fresh water} = \frac{\text{WPA}}{100} \times 1.000$$

$$\text{TPC}_{FW} = \frac{7278}{100} = 72.78.$$

The 'products for area' are multiplied by their distances from a convenient point (which is usually at either one end or the middle ordinate) or a special rule is used (see pages 27 and 28) if the C.G. of an area between two ordinates is required. The sum of the 'products for moment' so obtained is divided by the sum of the 'products for area' to give the distance of the geometrical centre from the chosen point.

In the diagram above it is assumed that moments are being taken about the end ordinate. The 'product for area' for this ordinate is thus multiplied by (0 × h) to get the 'product for moment'.

Simpson's rules – quadrature

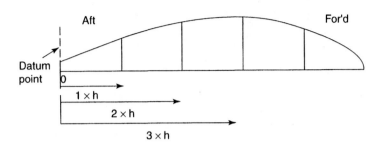

Figure 2.6

The distances for the succeeding 'products for area' will be (1 × h), (2 × h), (3 × h) and so on from the Datum. It will be seen that 'h' is a common factor and in order to make the arithmetic easier we may divide all our distances by it, thus making levers of 0, 1, 2, 3 and so on.

WORKED EXAMPLE 14

The 1/2-ordinates of a vessel's waterplane starting from forward and spaced 36 m apart are: 1.3, 11.2, 16.3, 17.5, 14.4, 8.7 and 3.0 metres respectively. Calculate the position of the longitudinal geometrical centre (LCF):

(a) about the fore end ⎫
⎬ use Simpson's first rule
(b) about amidships ⎭

(a)

	1/2-ord	SM	Area ftn	lever	Moment ftn
DATUM	1.3	1	1.3	0	0
	11.2	4	44.8	1	44.8
	16.3	2	32.6	2	65.2
✕	17.5	4	70.0	3	210.0
	14.4	2	28.8	4	115.2
	8.7	4	34.8	5	174.0
AFT	3.0	1	3.0	6	18.0
			215.3 = $\sum_1$		627.2 = $\sum_2$

$$\text{LCF from fore end} = \frac{\sum_2}{\sum_1} \times h$$

$$= \frac{627.2}{215.3} \times 36 = 104.87 \text{ m} \quad \text{(equivalent to 3.13 m for'd of ✕)}$$

(b)

	1/2-ord	SM	Area ftn	Lever	Moment ftn
FOR'D	1.3	1	1.3	+3	3.9
	11.2	4	44.8	+2	89.6
	16.3	2	32.6	+1	32.6
DATUM	17.5	4	70.0	0	0
	14.4	2	28.8	−1	−28.8
	8.7	4	34.8	−2	−69.6
AFT	3.0	1	3.0	−3	−9.0
			$215.3 = \sum_1$		$-18.7 = \sum_2$ (i.e. for'd)

$$\text{LCF from amidships} = \frac{\sum_2}{\sum_1} \times h = \frac{-18.7}{215.3} \times 36$$

$LCF = -3.13 \text{ m}$, or $3.13 \text{ m for'd of } \text{\ss}$ (as for answer to part (a)).

Note that in 14(b) levers AFT of datum are +ve and levers FOR'D of datum are −ve.

The use of the 3rd rule enables the geometrical centre of an area between two adjacent ordinates to be found if an additional ordinate is known. The multipliers 3, 10 and −1 are used for this purpose in order to find the moment of area.

To use: Determine the area under consideration using Simpson's 3rd Rule. Then:

To 3 times the ordinate bounding the area, add 10 times the middle ordinate, subtract the other given ordinate and multiply this result by $\frac{1}{24}$ of the square of the common interval.

So given semi-ordinates y, y_1, and y_2

Moment ABCD = $\frac{h^2}{24}(3y + 10y_1 - y_2) \times 2$

Distance of CG along PQ from P = x

$$x = \frac{\text{Moment}}{\text{Area}} = \frac{\sum_M}{\sum_A} \times \frac{h}{2}$$

Similarly:

Moment CDEF = $\frac{h^2}{24}(3y_2 + 10y_1 - y) \times 2$

Simpson's rules – quadrature

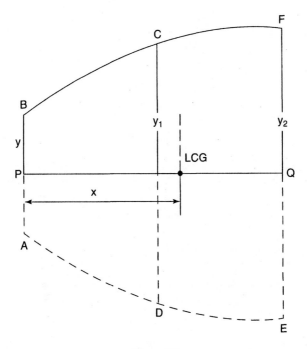

Figure 2.7

Distance of CG along QP from Q:

$$\text{LCG} = \frac{\text{Moment}}{\text{Area}}$$

$$\text{LCG} = \frac{h^2/24 \times \sum_M \times 2}{1/12 \times h \times \sum_A \times 2}$$

$$\therefore \text{LCG} = \frac{\sum_M}{\sum_A} \times \frac{h}{2} \text{ from Q.}$$

Worked Example 15 shows the method to be used. $\sum$ = 'summation of'.

WORKED EXAMPLE 15

The following 1/2-ordinates are spaced 20 m apart: 7.7, 12.5 and 17.3 m. Find the deck area between each ordinate and the longitudinal position of its geometrical centre for each area.

Area between 1st & 2nd ords		
1/2-ord	SM	Area ftn
7.7	5	38.5
12.5	8	100.0
17.3	−1	−17.3
		121.2 = $\sum_A$

Area between 2nd & 3rd ords		
1/2-ord	SM	Area ftn
7.7	−1	−7.7
12.5	8	100.0
17.3	5	86.5
		178.8 = $\sum_A$

$A_1 = 1/12 \times \sum_A \times h \times 2$

$A_1 = 1/12 \times 121.2 \times 20 \times 2$

$A_1 = 404 \, m^2$

$A_2 = 1/12 \times \sum_A \times h \times 2$

$A_2 = 1/12 \times 178.8 \times 20 \times 2$

$A_2 = 596 \, m^2$

TO FIND THE GEOMETRICAL CENTRES

1/2-ord	SM	Moment ftn
7.7	3	23.1
12.5	10	125.0
17.3	−1	−17.3
		130.8 = $\sum_M$

1/2-ord	SM	Moment ftn
7.7	−1	−7.7
12.5	10	125.0
17.3	3	51.9
		169.2 = $\sum_M$

Distance of CG from 7.7 ordinate

$$= \frac{\sum_M}{\sum_A} \times \frac{h}{2}$$

$$= \frac{130.8}{121.2} \times \frac{20}{2}$$

$$= 10.792 \, m.$$

Distance of CG from 17.3 ordinate

$$= \frac{\sum_M}{\sum_A} \times \frac{h}{2}$$

$$= \frac{169.2}{178.8} \times \frac{20}{2}$$

$$= 9.463 \, m.$$

WORKED EXAMPLE 16

The areas of equidistantly spaced vertical sections of the hull form below water-level are as shown. Length of ship is 400 m.

30, 226.4, 487.8, 731.6, 883.0, 825.5, 587.2, 262.1 and 39.8 square metres respectively.

Simpson's rules – quadrature

If the first vertical area (sectional area) was at the fore end, calculate this vessel's displacement in salt water and her longitudinal centre of buoyancy (LCB) from amidships.

	Section No.	Area	SM	Volume ftn	Lever	Moment ftn
for'd	1	30	1	30.0	−4	−120.0
	2	226.4	4	905.6	−3	−2716.8
	3	487.8	2	975.6	−2	−1951.2
	4	731.6	4	2926.4	−1	−2926.4
⋈	5	883.0	2	1766.0	0	0
	6	825.5	4	3302.0	+1	+3302.0
	7	587.2	2	1174.4	+2	+2348.8
	8	262.1	4	1048.4	+3	+3145.2
aft.	9	39.8	1	39.8	+4	+159.2
				$12\,168.2 = \sum_1$		$+1240.8 = \sum_2$

$\sum$ = 'summation of'. (i.e. in aft body)

$$\text{Displacement} = \frac{1}{3} \times \sum_1 \times h \times \rho_{SW} \qquad h = \frac{400}{8} = 50\,\text{m}$$

$$= \tfrac{1}{3} \times 12\,168.2 \times 50 \times 1.025$$

Displacement = 207 873 tonnes.

$$\text{LCB} = \frac{\sum_2}{\sum_1} \times h$$

$$= \frac{+1240.8}{12\,168.2} \times 50$$

LCB = +5.10 m i.e. 5.10 m AFT of amidships.

WORKED EXAMPLE 17

The following is an extract from a vessel's hydrostatic table of figures:

Draft	7 m	8 m	9 m	10 m	11 m	12 m	13 m	14 m
TPC	43.1	43.6	44.1	44.6	45.0	45.4	45.8	46.2

The displacement at a draft of 7 m is 15 000. Calculate the displacement at a draft of 14 m and the vertical position of centre of buoyancy (VCB) at this draft if the KB is 3.75 m at the 7 m draft.

Draft	TPC	SM		Volume ftn		Lever		Moment function	
14	46.2	1		46.2		4		184.8	
13	45.8	4		183.2		3		549.6	
12	45.4	2		90.8		2		181.6	
11	45.0	4		180.0		1		180.0	
10	44.6	1	1	44.6	44.6	0	3	0	133.8
9	44.1		3		132.3		2		264.6
8	43.6		3		130.8		1		130.8
7	43.1		1		43.1		0		0
				$544.8 = \sum_1$	$350.8 = \sum_2$			$1096.0 = \sum_3$	$529.2 = \sum_4$

$\sum$ = 'summation of'.

$$\text{Displacement 10 m to 14 m} = \frac{1}{3} \times \sum_1 \times h \times 100$$

$$= \tfrac{1}{3} \times 544.8 \times 1 \times 100 = 18\,160 \text{ tonnes.}$$

$$\text{Displacement 7 m to 10 m} = \frac{3}{8} \times \sum_2 \times h \times 100$$

$$= \tfrac{3}{8} \times 350.8 \times 1 \times 100 = 13\,155 \text{ tonnes.}$$

Displacement up to 14 m = $18\,160 + 13\,155$ + appendage of $15\,000\,\text{t} = 46\,315$ tonnes.

$$\text{TPC} = \frac{1.025 \times A}{100} \qquad \therefore A = \frac{100 \times \text{TPC}}{1.025}$$

and W = Vol of Δ't $\times$ 1.025, where Δ't = displacement.

$\therefore$ The density value of 1.025 cancels out top and bottom so all we need to do is use the *multiplier of 100* as shown.

$$\text{VCB above 10 m datum for upper portion} = \frac{\sum_3}{\sum_1} \times h = \frac{1096.0 \times 1}{544.8} = 2.0117\,\text{m}$$

$$= 12.0117\,\text{m above keel.}$$

$$\text{VCB above 7 m datum for lower portion} = \frac{\sum_4}{\sum_2} \times h = \frac{529.2 \times 1}{350.8} = 1.5086\,\text{m}$$

$$= 8.5086\,\text{m above keel.}$$

Now summarise using at moment table.

Item	Displacement	KB	Moment of Wgt
Upper portion	18 160	12.0117	218 312
Lower portion	13 155	8.5086	111 931
Appendage	15 000	3.7500	56 250
	46 315 = $\sum_5$		386 313 = $\sum_6$

Total displacement = $\sum_5$ = 46 315 tonnes.

$$KB = \frac{\sum_6}{\sum_5} = \frac{386\,313}{46\,315}$$
$$= 8.34 \text{ m}$$

So VCB = 8.34 m, above base, using Simpson's Rules.

Check by Morrish's formula:

$$KB = d - \frac{1}{3}\left(\frac{d}{2} + \frac{V}{A}\right)$$

$$KB = 14 - \frac{1}{3}\left(\frac{14}{2} + \frac{46\,315}{1.025} \times \frac{1.025}{46.2 \times 100}\right)$$

∴ KB = 8.33 m, which is very near to previously obtained value!!

WORKED EXAMPLE 18

From the following information, calculate the vessel's Deadweight, the fully-loaded displacement and C_D co-efft.

Light draft	8 m,	WPA = 9750 m².
Medium Ballast draft	10 m,	WPA = 11 278 m².
Heavy Ballast draft	12 m,	WPA = 12 600 m².
Fully-loaded draft	14 m,	WPA = 13 925 m².

Light weight @ 8 m draft is 18 231 tonnes.

Waterplane area	SM	Volume ftn
9 750	1	9 750
11 278	3	33 834
12 600	3	37 800
13 925	1	13 925
		95 309 = $\sum_2$

$\sum$ = 'summation of'. $\Delta't$ = displacement in tonnes.

Let Volume of displacement between 8 m and 14 m = V_1.

$$\therefore V_1 = \frac{3}{8} \times \sum_2 \times h$$

$$\therefore V_1 = \frac{3}{8} \times 95\,309 \times 2$$

$$\therefore V_1 = 71\,482\,m^3.$$

$$\therefore DWT = V_1 \times \rho_{SW}$$

$$= 71\,482 \times 1.025\,tonnes$$

$\therefore$ DWT = 73 269 tonnes
+ Lightweight (as given) = +18 231 tonnes
Fully-loaded displacement = 91 500 tonnes

$$C_D = \frac{DWT}{\Delta't} = \frac{73\,269}{91\,500} = 0.801.$$

DWT = 73 269 t, fully-loaded displacement = 91 500 t and C_D = 0.801.

SUB-DIVIDED COMMON INTERVALS
WORKED EXAMPLE 19

The 1/2-ords for part of an Upper deck are as follows

Station	6	7	8	$8\frac{1}{2}$	9	$9\frac{1}{4}$	$9\frac{1}{2}$	$9\frac{3}{4}$	10
1/2-ord (m)	8.61	8.01	7.02	6.32	5.32	4.74	3.84	2.72	0

If the ships LBP is 120 m, then calculate the area and LCG for'd of station 6 for this portion of deck.

Simpson's rules – quadrature

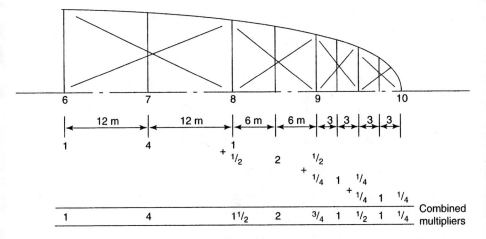

Figure 2.8

STN	1/2-ord	SM	Area ftn	Lever$_{STN\ 6}$	Moment ftn
6	8.61	1	8.61	0	–
7	8.01	4	32.04	1	32.04
8	7.02	$1\frac{1}{2}$	10.53	2	21.06
$8\frac{1}{2}$	6.32	2	12.64	$2\frac{1}{2}$	31.60
9	5.32	$\frac{3}{4}$	3.99	3	11.97
$9\frac{1}{4}$	4.74	1	4.74	$3\frac{1}{4}$	15.41
$9\frac{1}{2}$	3.84	$\frac{1}{2}$	1.92	$3\frac{1}{2}$	6.72
$9\frac{3}{4}$	2.72	1	2.72	$3\frac{3}{4}$	9.18
10	0	$\frac{1}{4}$	–	4	–
			$77.19 = \sum_1$		$127.98 = \sum_2$

$\sum$ = 'summation of'.

$h = \dfrac{LBP}{10} = \dfrac{120}{10} = 12\,m$

$$\text{Area} = \frac{1}{3} \times h \times \sum\nolimits_1 \times 2 = \frac{1}{3} \times 12 \times 77.19 \times 2 = 617.5\,m^2.$$

$$\text{LCG for'd of Station 6} = \frac{\sum_2}{\sum_1} \times h = \frac{127.98}{77.19} \times 12 = 19.9\,m.$$

Note:

If we halve the interval, we halve the multipliers to $\frac{1}{2}$, 2, $\frac{1}{2}$.

If we quarter the interval, we quarter the multipliers to $\frac{1}{4}$, 1, $\frac{1}{4}$.

Always use the LARGEST common interval in the final calculation i.e. the h of 12 m.

Remember that $\frac{1}{2}$-ords are also known as offsets.

MOMENTS OF INERTIA (SECOND MOMENTS OF AREA), USING SIMPSON'S RULES

WORKED EXAMPLE 20

The half-ordinates for a ships waterplane at equidistant intervals from aft are as follows:

1.9, 5.3, 8.3, 9.8, 9.8, 8.3, 5.2, 1.3 and 0 metres.

(a) If the common interval was 15.9 m, then calculate the second moment of area about (i) amidships ($I_{\bowtie}$) and (ii) about long'l centre of flotation (I_{LCF}).

(b) If the volume of displacement for the ship is 10 000 m³, proceed then to estimate long'l BM.

STN	1/2-ord	SM	Area ftn	$L_{\bowtie}$	Moment ftn	$L_{\bowtie}$	Inertia$_{\bowtie}$ ftn
AP	1.9	1	1.9	+4	+7.6	+4	30.4
	5.3	4	21.2	+3	+63.6	+3	190.8
	8.3	2	18.6	+2	+37.2	+2	74.4
	9.8	4	19.6	+1	+19.6	+1	19.6
Datum ⋈	9.8	2	19.6	0	0	0	0
	8.3	4	33.2	−1	−33.2	−1	33.2
	5.2	2	10.4	−2	−21.8	−2	43.6
	1.3	4	5.2	−3	−15.6	−3	46.8
FP	0	1	0	−4	−	−4	−
			129.7 = $\sum_1$		+57.4 = $\sum_2$		438.8 = $\sum_3$

$\sum$ = 'summation of'.

$$\text{WPA} = \frac{1}{3} \times \sum_1 \times \text{CI} \times 2 = \frac{1}{3} \times 129.7 \times 15.9 \times 2 = 1375 \, \text{m}^2.$$

$$\text{LCF} = \frac{\sum_2}{\sum_1} \times \text{CI} = \frac{57.4}{129.7} \times 15.9 = 7.04 \, \text{m AFT of } \bowtie.$$

$$I_{\bowtie} = \frac{1}{3} \times \sum_3 \times \text{CI}^3 \times 2 = \frac{1}{3} \times 438.8 \times 15.9^3 \times 2$$

$$\therefore I_{\bowtie} = 1\,175\,890 \, \text{m}^4$$
$$-Ay^2 = -1375 \times 7.04^2 = \underline{68\,147 \, \text{m}^4}$$
$$I_{LCF} = \underline{1\,107\,743 \, \text{m}^4}$$

$$BM_L = \frac{I_{LCF}}{Vol\ of\ \Delta't} = \frac{1\,107\,743}{10\,000} = 110.78\,m.$$

In the Worked Example 20, the parallel axis theorem was used. To explain further:

In *applied mechanics*; $I_{NA} = I_{XX} - Ay^2$ ... parallel axis theorem

In *ship technology*; $I_{LCF} = I_{⊗} - WPA(LCF_{⊗})^2$.

I_{NA} and I_{LCF} represent Moments of Inertia (longitudinal) about the LCG of the considered area.

$$\text{Hence } I_{LCF} = I_{⊗} - 1375 \times 7.04^2$$
$$= 1\,175\,890 - 68\,147$$
$$= 1\,107\,743\,m^4.$$

WORKED EXAMPLE 21

(a) Calculate the Moment of Inertia about the centreline $I_{\mathcal{C}}$ for the waterplane in the previous question.

(b) Proceed to evaluate the transverse BM if again the volume of displacement is $10\,000\,m^3$.

STN	1/2-ords	$(1/2 - ords)^3$	SM	$I_{\mathcal{C}}$ function
AP	1.9	7	1	7
	5.3	149	4	596
	8.3	572	2	1144
	9.8	941	4	3764
⊗	9.8	941	2	1882
	8.3	572	4	2288
	5.2	141	2	282
	1.3	2	4	8
FP	0	0	1	0
				$9971 = \sum_1$

$\sum$ = 'summation of'.

$$I_{\mathcal{C}} = \frac{1}{9} \times \sum_1 \times CI \times 2 = \frac{1}{9} \times 9971 \times 15.9 \times 2 = 35\,231\,m^4$$

$$BM_T = \frac{I_{\mathcal{C}}}{V} = \frac{35\,231}{10\,000} = 3.52\,m.$$

Note how $I_{\mathcal{C}}$ is *much smaller* then I_{LCF} and $I_{⊗}$ and BM_T is *much smaller* than BM_L.

When using Simpson's Rules, where possible the steps for procedure should be: Sketch, Table, Calculations.

CHAPTER THREE

Bending of Beams and Ships

In Chapter one it was shown that the total weight of the ship and contents is equal to the total upthrust due to buoyancy if the vessel is to float. However, although the weight is considered to act vertically downwards through the centre of gravity, this total is composed of many individual weights which act at many different parts of the ship.

The total upthrust due to buoyancy is considered to act vertically upwards through the centre of buoyancy but this total upthrust is composed of countless individual upthrusts acting on the plating of the ship.

If the downward force due to weight at a point and the upward thrust due to buoyancy at the same point are not equal, stresses will occur. These stresses can be expressed as Shear Forces and Bending Moments. Steps taken to counteract their effects are detailed in the revised companion volume *Ship Construction Sketches and Notes* (1997 Edition).

Whilst finding these shear forces and bending moments in practice is a somewhat complicated exercise, the theory is similar to that of shear forces and bending moments on a beam, and these are first considered.

SHEAR FORCE

When a section such as a beam is carrying a load there is a tendency for some parts to be pushed upwards and for other parts to move downwards, this tendency is termed shearing. The shear force at a point or station is the vertical force at that point. The shear force at a station may also be defined as being the total load on either the left hand side or the right hand side of the station: load being defined as the difference between the downward and upward forces.

Consider first of all the following beam theory for simply supported beams and for cantilever structures.

If the beam in Fig. 3.1 is static and supported at its ends, the total forces upwards (reaction at the pivots) must equal the downward forces (loads).

$$R_1 + R_2 = W \qquad (1)$$

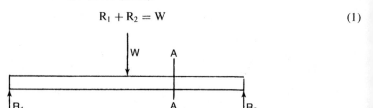

Figure 3.1

To the left of the line AA there is a resultant downward force $W - R_1$.

To the right of the line AA there is an upward force R_2.

The shear force at a point on the beam along AA is either $W - R_1$ or R_2.

It can be seen from (1) above that these two quantities are the same.

Bending Moment

The beam which we have been considering would also have a tendency to bend and the bending moment measures this tendency. Its size depends on the amount of the load as well as how the load is placed together with the method of support. Bending moments are calculated in the same way as ordinary moments i.e. multiplying force by distance, and so they are expressed in weight-length units. As with the calculation of shear force the bending moment at a station is obtained by considering moments **either** to the left **or** the right of the station.

In the diagrams on subsequent pages shear forces and bending moments are drawn in accordance with the understated sign convention. 'If a *downward* force (weight loaded) is considered *negative* and a reaction or *upthrust* considered *positive* then the shear force is measured upwards from the zero line if loads to the left of the station result in a positive shear force, or if loads to the right of the station result in a negative shear force. The force is measured downward from the zero line if the opposite signs result'. A positive shear is illustrated for point X in Fig. 3.2(a). 'Bending moments will likewise be positive or negative and a positive moment resulting from consideration of loads **either** to the left **or** to the right of the station is measured below the zero line and a negative moment resulting from consideration of loads is measured above the line'. Positive moments cause sagging and are illustrated in Fig. 3.2(b).

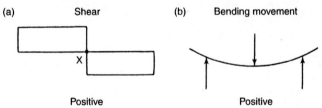

Figure 3.2

In the following theory it is sometimes convenient to consider a point as being an infinitesimal distance to the left or right of the station at which the shear force or bending moment is required. Such a position is suffixed with the letter L or R.

A HORIZONTAL BEAM WITH END SUPPORTS

a) Weightless with a point load at the centre

If stationary, $W = R_a + R_b$

To find R_a and R_b.

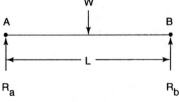

Figure 3.3(a)

Taking moments about A

Anti-clockwise moments = Clockwise moments

$$R_b \times L = W \times \frac{L}{2}$$

$$R_b = \frac{W}{2}$$

If pivot B supports half the weight the rest must be supported at A.

Therefore $R_a = \frac{W}{2}$

Station	Shear Force	Bending Moment
A_R	R_a or $\frac{W}{2}$	0
1/4 L	R_a or $\frac{W}{2}$	$\frac{W}{2} \times \frac{L}{4} = \frac{WL}{8}$
1/2 L_L	R_a or $\frac{W}{2}$	$\frac{W}{2} \times \frac{L}{2} = \frac{WL}{4}$
1/2 L_R	$R_a - W = -\frac{W}{2}$	$\frac{W}{2} \times \frac{L}{2} = \frac{WL}{4}$

It may be noted that the shear forces and bending moments shown above have been obtained by considering loads and moments to the left of the stations. Exactly the same results would have been obtained if loads and moments to the right of the stations had been considered, these are shown below. The reason for the different signs in the shear forces will be apparent from reference to the text on page 37.

Station	Shear Force	Bending Moment
A	$R_b - W = -\frac{W}{2}$	$\left(\frac{W}{2} \times L\right) - \left(W \times \frac{L}{2}\right) = 0$
1/4 L	$R_b - W = \frac{W}{2}$	$\left(\frac{W}{2} \times \frac{3L}{4}\right) - \left(W \times \frac{L}{4}\right) = \frac{WL}{8}$
1/2 L_L	$R_b - W = -\frac{W}{2}$	$\frac{W}{2} \times \frac{L}{2} = \frac{WL}{4}$
1/2 L_R	$R_b = \frac{W}{2}$	$\frac{W}{2} \times \frac{L}{2} = \frac{WL}{4}$

The foregoing theory should emphasise that the shear force at a point is found by the algebraic summing of the loads either to the left or the right of that point. Likewise the bending moment at a station is the algebraic sum of the moments either to the left or the right of the station.

If the convention shown on page 37 is used the shear forces and bending moments could be plotted on a diagram as shown below.

NOTE: Both the tables on page 38 and the Fig. 3.3(b) show that the area under the shear force curve up to a point is equal to the bending moment at that point.

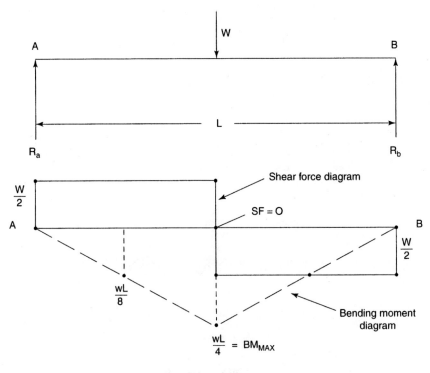

Figure 3.3(b)

b) Weightless with an evenly spread load. This has exactly the same effect as a beam of uniform section whose weight is distributed uniformly along the beam. See Figure 3.4(a)

If the total load is W and the length is L, then the weight per unit length w equals $\dfrac{W}{L}$.

Reaction $R_a = R_b = \dfrac{W}{2}$

Station	Shear Force	Bending Moment
A_R	$R_a = \dfrac{W}{2}$	0
1/4 L	$R_a - \dfrac{wL}{4}$	$\dfrac{W}{2} \times \dfrac{L}{4} - \dfrac{wL}{4} \times \dfrac{L}{8}$
	or $\dfrac{W}{2} - \dfrac{W}{4} = \dfrac{W}{4}$	or $\dfrac{WL}{8} - \dfrac{WL}{32} = \dfrac{3WL}{32}$
1/2 L	$R_a - \dfrac{wL}{2}$	$\dfrac{W}{2} \times \dfrac{L}{2} - \dfrac{wL}{2} \times \dfrac{L}{4}$
	or $\dfrac{W}{2} - \dfrac{W}{2} = 0$	or $\dfrac{WL}{4} - \dfrac{WL}{8} = \dfrac{WL}{8}$
B	$R_a - wL$	$\dfrac{W}{2} \times L - wL \times \dfrac{L}{2}$
	or $\dfrac{W}{2} - W = \dfrac{W}{2}$	or $\dfrac{WL}{2} - \dfrac{WL}{2} = 0$

The above have been obtained by considering loads and bending moments to the left of each station. It is suggested that the reader now considers the loads and bending moments to the right of each station. The values of the shear forces and bending moments so found will be exactly the same as those above.

The shear force and bending moment diagram shown in Fig. 3.4(b) will be typical of any uniformly loaded beam.

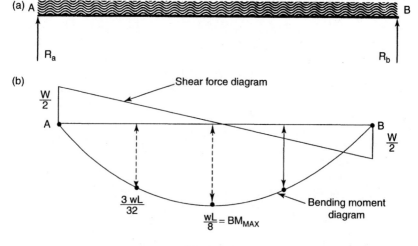

Figure 3.4

c) A beam having an unevenly spread load.

In this case a curve of loads would be drawn, each ordinate representing the average load per unit length at that part of the beam. The general principles already described can then be followed namely that the shear force at any point of the beam is the resultant of the upthrusts and loads on one side of that point. The bending moment at any point can be obtained by finding the area under the shear force curve up to that point. Worked Example 23 shows how this is done in the case of a ship.

WORKED EXAMPLE 22

A uniform beam AB 6 metres in length and weight 3 tonnes is supported at its ends. Weights of 1 tonne and 2 tonnes are loaded at points 2 metres and 5 metres from the end A. Calculate the shear force and the bending moment at 1 m intervals along this beam from A. What are the values of SF_{MAX}, SF_{MIN} and BM_{MAX}?

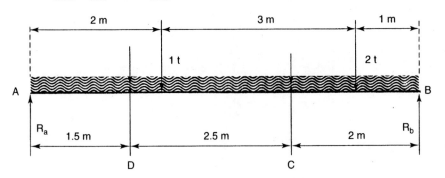

Figure 3.5

Weight of the beam is 3 tonnes or 0.5 tonne per metre.

To find the reactions at A and B:

The total moment about any point caused by the reactions

$$= \text{The total moment about that point caused by the loads.}$$

Taking moments about A:

$(R_a \times 0) + (R_b \times 6) = (3 \times 3) + (1 \times 2) + (2 \times 5) \quad \therefore 6 \cdot R_b = 21$

$$\therefore R_b = \frac{21}{6} = 3.5 \text{ tonnes}$$

$R_a + R_b \qquad = 6.0 \text{ tonnes}$

$R_a \qquad\qquad = 2.5 \text{ tonnes}$

Calculation of Shear Forces @ 1 m intervals along the beam.

$$SF_0 = 0 \text{ and } +2.5\,t$$
$$SF_1 = 2.5 - 0.5 = +2\,t$$
$$SF_2 = 2.5 - 1 = 1.5\,t \text{ and } 0.5\,t$$
$$SF_3 = 2.5 - 1.5 - 1 = 0\,t \text{ @ mid-length.}$$
$$SF_4 = 0 - 0.5 = -0.5\,t$$
$$SF_5 = -0.5 - 0.5 = -1\,t \text{ and } -3\,t$$
$$SF_6 = 2.5 + 3.5 - 1 - 2 - 3 = 0 \text{ and } -3.5\,t$$

Calculation of Bending Moments

$$BM_0 = \text{zero t.m.}$$
$$BM_{(1/2)} = \left(\frac{2(1/2) + 2(1/4)}{2}\right) \times 1/2 = 1.1875\,\text{t.m.}$$
$$BM_1 = \left(\frac{2.5 + 2}{2}\right) \times 1 = 2.25\,\text{t.m.}$$
$$BM_2 = \left(\frac{2.5 + 1.5}{2}\right) \times 2 = 4\,\text{t.m.}$$
$$BM_3 = 4 + \left(\frac{0.5 + 0}{2}\right) \times 1 = 4.25\,\text{t.m.}$$
$$BM_4 = 4.25 - \left(\frac{0.5 + 0}{2}\right) \times 1 = 4\,\text{t.m.}$$
$$BM_5 = 4.25 - \left(\frac{0 + 1}{2}\right) \times 2 = 3.25\,\text{t.m.}$$
$$BM_{5(1/2)} = 0 - \left(\frac{3(1/4) + 3(1/2)}{2}\right) \times 1/2 = 1.6875\,\text{t.m.}$$
$$BM_6 = 3.25 - \left(\frac{3 + 3.5}{2}\right) \times 1 = \text{zero t.m.}$$

Fig. 3.6 shows a graphical plot of the above SF and BM values along this 6 m beam.

It can be seen from an analysis of Fig. 3.6 that it contains three features:

(i) The greatest Bending Moment occurs at midlength.
(ii) There is a sharp discontinuity at the points where there are concentrated loads. This is similar to the shear curve for the weightless beam with point loading.

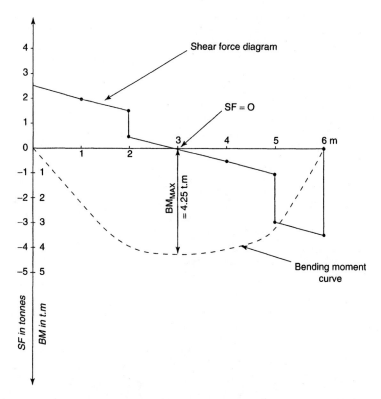

Figure 3.6 SF and BM diagrams for the 6 m beam. $SF_{max} = 2.5\,t$, $SF_{min} = -3.5\,t$, $BM_{max} = 4.25\,t.m$.

(iii) There is a gradual slope where the load is uniformly distributed. This is similar to the weightless beam with a uniform load.

The zero value of the shear curve is occurs at the position of maximum bending moment at midlength.

So far beams supported at each end have been considered. If a beam is not supported at the extreme ends different problems arise in calculating shear forces and bending moments beyond the points of support although the general theory is the same.

The part of the beam beyond the end support may be considered as a cantilever which is a beam with one end fixed and the other end free. Shear forces and bending moments for two conditions of loading a cantilever follow.

When calculating shear forces and bending moments at a station in the case of a cantilever beam the loads to be considered should be those from the station towards the free end of the beam. If the free end is always considered as being to the right then the sign convention on page 37 is applicable.

(a) Weightless with point loading at the free end.

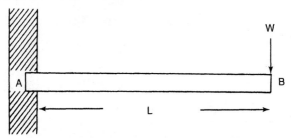

Figure 3.7

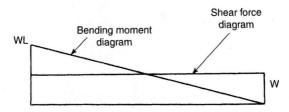

Figure 3.8

Station	Shear force	Bending Moment
B_L	$-W$	0
$1/2L$	$-W$	$-\dfrac{WL}{2}$
A	$-W$	$-WL$

(b) Weightless with evenly spread load.

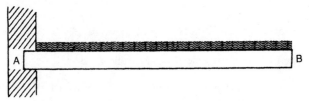

Figure 3.9

Weight per metre run w equals $\dfrac{W}{L}$

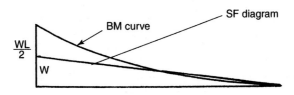

Figure 3.10

Station	Shear Force	Bending Moment
B	0	0
1/2 L	$-\dfrac{W}{2}$	$-\dfrac{wL}{2} \times \dfrac{L}{4} = -\dfrac{WL}{8}$
A	$-W$	$-wL \times \dfrac{L}{2} = -\dfrac{WL}{2}$

APPLICATION TO SHIPS

As ships may be considered as a form of beam the theory so far discussed can be applied. The loads along the ship are found by first plotting a curve of weights, each ordinate representing the average weight per unit length at that point. The weight per unit length would vary dependent upon the loading – the greater the weight the shorter the distance. A second curve is then drawn, this is the curve of upthrusts or buoyancy.

Each of the curves can vary, the curve of weights is dependent on the type of loading (light or loaded conditions for example). The curve of buoyancy could be drawn for either the still water condition or for when the vessel is supported by waves at the ends (sagging) or when she is supported by a wave at amidships (hogging).

The difference between the curves of weight and buoyancy at any point is the load which is drawn as an ordinate at that point. Joining each ordinate gives the curve of loads and from the information on this the shear force and thence the bending moments can be found. The worked example which follows illustrates this.

WORKED EXAMPLE 23

A box shaped barge, length 30 m, breadth 8 m and depth 6 m, floats at an even keel draft to 4 m in saltwater of relative density 1.025, has 500 tonnes of ore spread over the midship half-length and two cases of machinery each weighing 20 tonnes, measuring 2 m × 2 m × 2 m, stowed on the centre line 5 m from each end.

Construct a curve of loads for the still water condition assuming the weight of the barge to be evenly distributed over the full length and from it draw curves for Shear Force and Bending Moments.

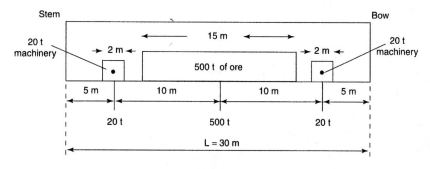

Figure 3.11

$$\text{Displacement of barge} = 30 \times 8 \times 4 \times 1.025$$
$$= 984 \text{ tonnes}$$
$$\text{Weight of cargo} = -540 \text{ tonnes}$$
$$\text{Weight of barge} = \overline{444} \text{ tonnes}$$

$$\text{Weight of barge per metre run} = \frac{444}{30} = 14.8 \text{ tonnes}$$

$$\text{Buoyancy per metre run} = \frac{984}{30} = 32.8 \text{ tonnes}$$

$$\text{Ore spread per metre run} = \frac{500}{15} = 33.3 \text{ tonnes}$$

$$\text{Machinery spread run} = \frac{20}{2} = 10.0 \text{ tonnes}$$

Keeping to the convention on page 37 the buoyancy per metre run is laid off upwards from the datum line AB and is represented by rectangle ABCD. Weight reduces the effect of buoyancy so that it is laid off downwards from line CD. The weight per metre run of the ship is represented by CE and the weight of the ship by rectangle CDEF. The weights of the cargo will further reduce the effect of buoyancy and these are laid downwards from EF in their appropriate fore and aft positions being represented by rectangles GIJH, KMNL and OQRP. The portion of the buoyancy curve which is intact above AB represents positive loads and the weight curve below AB represents negative loads. The load curve is now drawn and is shown as a dotted line in Fig. 3.12.

The method outlined above is very suitable for the still water condition where buoyancy is evenly distributed. For such a vessel in a seaway, and in more complicated work, both buoyancy and weight are plotted from the baseline AB.

Summing the loads to the left or right at the various stations gives the shear force at those points. The summing of the loads is done by obtaining the area under the curve of loads up to the station, as the ordinates of the load curve are weights per metre run, if these are multiplied by the distance between them a load is obtained. This load is in weight units as

$$\frac{W}{L} \times L = W$$

Bending of beams and ships

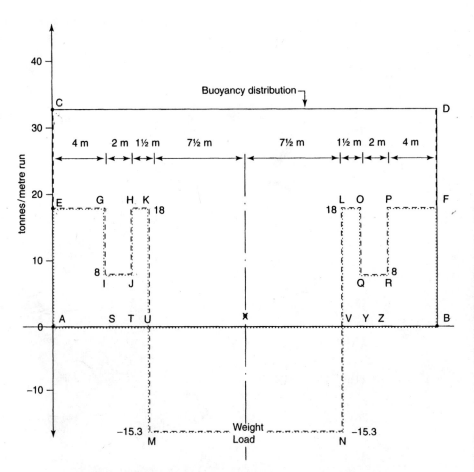

Figure 3.12 Ship Strength Diagrams ~ Weight Buoyancy and Load.

Station letter	From Point A	Area under load curve to left of station =	Shear force in tonnes
A	0 m	0	0
S	4 m	(4 × 18)	72
T	6 m	72 + (2 × 8)	88
U	7(1/2) m	88 + (1.5 × 18)	115
X	15 m	115 − (7.5 × 15.3)	0
V	22(1/2) m	0 − (7.5 × 15.3)	−115
Y	24 m	−115 + (1.5 × 18)	−88
Z	26 m	−88 + (2 × 8)	−72
B	30 m	−72 + (4 × 18)	0

If areas had been taken to the right of the various stations the same numerical results would have been obtained but with opposite signs as per the sign convention. The reader may care to check that this is so. The shear force curve is now drawn with the above values. The area under this latter curve up to the various stations gives the bending moment at those stations as follows:

Station letter	From Point A	Area under shear curve to left of the station	=	Bending Moment in tonnes metres
A	0 m	0		0
S	4 m	$4\left(\dfrac{0+72}{2}\right)$		144
T	6 m	$144 + 2\left(\dfrac{72+88}{2}\right)$		304
U	7(1/2) m	$304 + 1.5\left(\dfrac{88+115}{2}\right)$		456.25
X	15 m	$456.25 + 7.5\left(\dfrac{115+0}{2}\right)$		887.50
V	22(1/2) m	$887.5 - 7.5\left(\dfrac{0+115}{2}\right)$		456.25
Y	24 m	$456.25 - 1.5\left(\dfrac{115+88}{2}\right)$		304
Z	26 m	$304 - 2\left(\dfrac{88+72}{2}\right)$		144
B	30 m	$144 - 4\left(\dfrac{72+0}{2}\right)$		0

The same results would have been obtained by taking moments to the right of the stations.

Figure 3.13 is a graphical representation of the SF and BM values calculated in the last two tables.

Let f_{max} = maximum bending stress in t/m²

Then $f_{MAX} = \dfrac{M}{I_{NA}} \times y$.

where M = BM_{MAX} obtained from ships BM curve in t.m.
y = vertical distance from the Neutral Axis to the plating under consideration in metres.
I_{NA} = moment of Inertia about the ships Neutral Axis in m⁴.

Bending of beams and ships

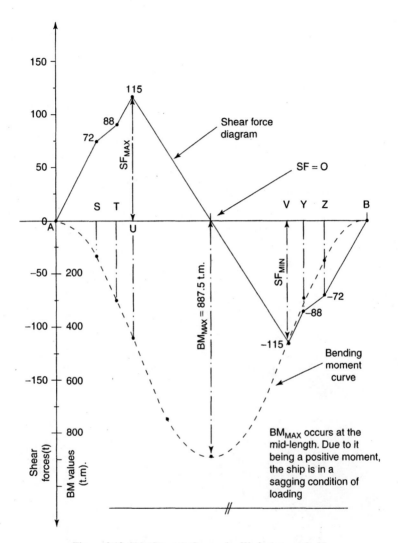

Figure 3.13 Ship Strength Curves for Worked example 23.

In Figure 3.14, ship was in a sagging condition of loading in a sea-way. Bottom shell fractured first followed by the ship breaking her back. Observe the Bulbous Bow in Figure 3.14.

Bulbous Bows are fitted on ships because they can:

a) give an increase in speed for similar input of Engine power. This may be up to $+1/2$ kt when fully-loaded and up to $+3/4$ kt in ballast condition.

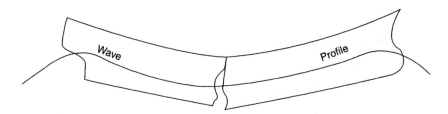

Figure 3.14 An example of poor mechanical ship stability.

b) give extra strength at bottom of Fore Peak tank.

c) reduce vibration amplitudes in the Fore Peak tank.

However, Bulbous Bows are expensive to install. On small, slow service speed vessels they can actually cause increased resistance to forward motion.

CHAPTER FOUR

Transverse Stability (Part 1)

Ship Stability depends on KB, BM, KG and GM. See diagrams on page xvii at beginning of the book. As will be later explained of these four values, GM is the most important.

Metacentre is at the intersection of vertical lines through the centres of buoyancy in the initial and slightly inclined positions.

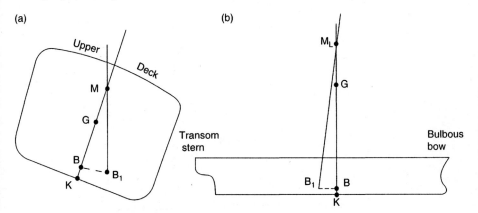

Figure 4.1

The transverse metacentre (M) is used when transverse inclinations (heel and list) are considered.

The longitudinal metacentre (M_L) is used when longitudinal inclinations (trim) are considered.

Metacentric Height (GM) is the distance between the centre of gravity and the metacentre.

The position of the metacentre may be calculated from the formula.

$$BM = \frac{I}{V}$$ the proof of which is shown in the following text:

where I is the moment of inertia of the waterplane in metre^4 units
and   V is the volume of displacement in cubic metres.

Note: The moment of inertia, which is also called the second moment of area, of a body about a line can be obtained by multiplying the mass of each and every particle in the body by the square of its distance from the line about which the moment of inertia is required. Adding the quantity found for each and every particle in the body gives the moment of inertia of the body.

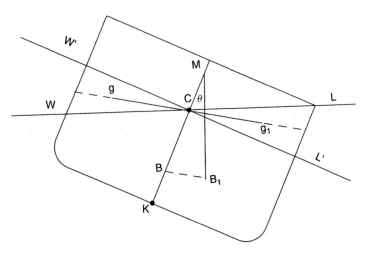

Figure 4.2

To prove $BM = \dfrac{I}{V}$

Wedge of immersion = Wedge of emersion

Let v = volume of either wedge of unit length

b = breadth of either wedge = $\tfrac{1}{2}$ breadth of ship

g and g_1 = centres of gravity of wedges

V = volume of displacement of ship

then $BB_1 = \dfrac{v \times gg_1}{V}$

Let the angle of heel θ be infinitely small so it can be said that WC, W¹C, LC, L¹C are all equal in length and triangles WCW¹ and LCL¹ are isosceles.

Now area of triangle $WCW^1 = \tfrac{1}{2}b^2 \sin\theta$ Note: $WW^1 = b\sin\theta$

Distance $gg_1 = \tfrac{4}{3}CL^1$ or $\tfrac{4}{3}b$

The moment of area $= \dfrac{b^2}{2}\sin\theta \times \dfrac{4}{3}b = \dfrac{2b^3}{3}\sin\theta$

If we consider a triangle of infinitely small length which we will call unity (1) then this moment is also the moment of the volume of the wedge at this particular point only. We must take the volume of all the other wedges and sum them up by integration or putting them through Simpson's Rules, the summation is denoted by $\sum$.

Transverse stability (part 1)

$$\text{Moment of whole wedge} = v \times gg_1 = \sum \frac{2b^3}{3} \sin\theta$$

θ being very small $\dfrac{BB_1}{BM} = \sin\theta$

$$BB_1 = BM \sin\theta$$

$$\text{Now } BB_1 = \frac{v \times gg_1}{V} = \frac{\sum \frac{2}{3}b^3 \sin\theta}{V}$$

$$BM \sin\theta = \frac{\sum \frac{2}{3}b^3 \sin\theta}{V}$$

$$BM = \frac{\sum \frac{2}{3}b^3 \sin\theta}{\sin\theta \times V} = \frac{\sum \frac{2}{3}b^3}{V}$$

But $\sum \frac{2}{3}b^3$ is, from the above work and by definition, the moment of inertia of the waterplane (I).

$$\therefore BM = \frac{I}{V}$$

If the BM's are known a curve of metacentres can be drawn by plotting metacentric heights against drafts an example for a box-shaped vessel follows. In a box shape I and V can be readily calculated as

I for a rectangular waterplane is $\dfrac{LB^3}{12}$ for *transverse* inclinations

or $\dfrac{BL^3}{12}$ for *longitudinal* inclinations.

The volume of displacement of a boxshape is $L \times B \times d$

Hence for *boxshape* $BM = \dfrac{LB^3}{12 \times L \times B \times d} = \dfrac{B^2}{12d}$ (*transversely*)

or $\dfrac{L^2}{12d}$ (*longitudinally*).

For a *triangular* shape $BM = \dfrac{B^2}{6d}$ (*transversely*)

or $\dfrac{L^2}{6d}$ (*longitudinally*).

For a *ship-shape* vessel:

$$BM = \frac{C_w^2 \times B^2}{12 \times d \times C_b} \quad (\text{approx'n})$$

and
$$BM_L = \frac{3 \times C_w^2 \times L^2}{40 \times d \times C_b} \quad (\text{approx'n}).$$

In all cases: L = the length of waterplane, B = maximum breadth of waterplane, d = draft, C_b = block co-efficient, C_w = waterplane area co-efft.

WORKED EXAMPLE 24

Draw a Metacentric Diagram for a boxshaped vessel 270 m in length, 24 m in breadth, for draft 2 m to 14 m at intervals of 1 m.

Draft (m)	BM (m)	KB (m)	KM (m)
2	24	1.0	25.0
3	16	1.5	17.5
4	12	2.0	14.0
5	9.6	2.5	12.1
6	8.0	3.0	11.0
7	6.8	3.5	10.3
8	6.0	4.0	10.0
9	5.3	4.5	9.8
10	4.8	5.0	9.8
11	4.4	5.5	9.9
12	4.0	6.0	10.0
13	3.7	6.5	10.2
14	3.4	7.0	10.4

The KB's and KM's are now plotted against drafts to form a Metacentric Diagram (Fig 4.3). For all types of ships, KB and KM values depend upon the underwater *geometrical form* of the vessel. KG on the other had depends on the *loading characteristics* of the ship.

In the above table of values:
$$KB = \frac{d}{2} \quad \text{and} \quad BM = \frac{B^2}{12 \times d}.$$
$$\text{So } BM = \frac{24 \times 24}{12 \times d} = \frac{48}{d} \text{ metres.}$$
$$KM = KB + BM \text{ for all drafts.}$$

USING A METACENTRIC DIAGRAM

(a) Assume that when draft is 5.5 m, KG is 8.48 m. This KG is then plotted as shown at point G_1. See Figure 4.3. KM measures 11.48 m so GM = 11.48 − 8.48 = +3 m.

GM is +ve, so ship is in stable condition.

(b) Assume that when draft is 10.5 m, KG is 11.82 m. This KG is then plotted as shown in Figure 4.3 at point G_2. KM measures 9.82 m so GM = 9.82 − 11.82 = −2 m.

GM is −ve, so ship is in unstable condition.

(c) Assume that when draft is 9.6 m, KG is 9.8 m. This KG is then plotted as shown in Figure 4.3 at G_3. KM measures 9.8 m, so GM = 9.8 − 9.8 = zero.

GM = 0, so ship is in neutral condition.

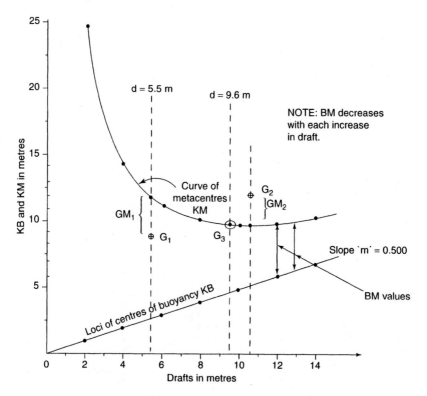

Figure 4.3 Metacentric Diagram for a box-shaped vessel.

For a ship-shape vessel, the Metacentric Diagram looks similar to that for the box-shaped vessel in Fig 4.3.

WORKED EXAMPLE 25

(a) Draw a Metacentric Diagram for the triangular-shaped vessel shown in Figure 4.4 for drafts up to 15 m.

(b) When draft is 13.5 m, KG is 10.5 m. Use the Metacentric Diagram to evaluate GM. Is this condition of loading stable or unstable?

(c) When draft is 7.5 m, KG is 7.55 m. Use the Metacentric Diagram to evaluate GM. Is this condition of loading stable or unstable?

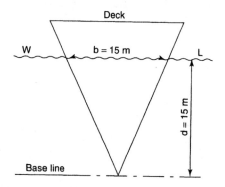

Draft	Br.Mld	$KB = \frac{2}{3}d$	$BM = \frac{B^2}{6d}$	KM
15	15	10	2.5	12.5
12	12	8	2.0	10.0
9	9	6	1.5	7.5
6	6	4	1.0	5.0
3	3	2	0.5	2.5

Figure 4.4 KB + BM = KM for all drafts.

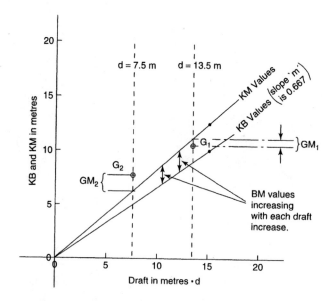

Figure 4.5 Metacentric Diagram for a triangular-shaped vessel.

(b) When draft = 13.5 m, KM = 11.25 m; If KG = 10.50 m, marked G_1 in Figure 4.5, then GM = 11.25 − 10.50 = +0.75 m, shown as GM_1.

GM is +ve, so ship is stable.

(c) When draft = 7.5 m, KM = 6.25 m; IF KG = 7.55 m marked G_2 in Figure 4.5, then GM = 6.25 − 7.55 = −1.30 m, shown as GM_2.

GM is −ve, so ship is unstable!!

Observations regarding the Metacentric Diagram for a triangular-shaped vessel as shown in Fig 4.5.

1. Waterline *breadths decrease* at lower drafts.
2. KB and KM are *straight* lines.
3. For KB line, the *slope 'm'* is 2/3 i.e. 0.667.
4. BM *increases* with increase in draft.
5. KB and KM are each *independent* of the *ship's length*.
6. KB and BM depend or geometrical hull-form of the ship.

Summary for Metacentric Diagrams:

Type of vessel	KB line 'm'	KM shape	BM change @ increasing draft
box-shaped	0.500	parabolic	decrease
ship-shaped	0.535 approx.	parabolic	decrease
triangular-shaped	0.667	straight	increase

Couples and Moments:

A couple is formed when two equal parallel forces are acting in opposite directions.

The lever of a couple is the perpendicular distance between the forces forming the couple.

Moment of a couple is the product of one of the forces forming the couple and the lever of the couple.

It has already been shown that the vessel's weight and the force of buoyancy must be equal for the vessel to float. If these forces are not on the same vertical line they will form a couple. Such a case is shown in Figure 4.6, a righting couple being formed when the vessel is heeled by the external force. The lever of the couple is known as the GZ or Righting lever.

Stability or statical stability is the ability of a vessel to return to her initial position after being forcibly inclined.

Moment of statical stability or righting moment is a measure of the vessel's ability to return to her initial position. It is always W × GZ tonnes-metres.

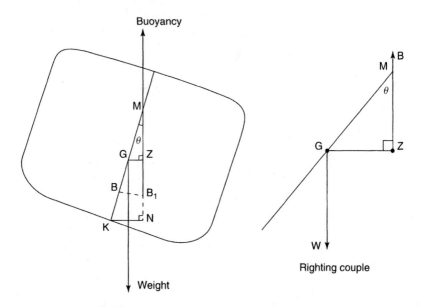

Figure 4.6

Metacentric stability. With this the metacentre is considered as being a fixed point. The GZ lever can then be expressed in terms of the metacentric height, i.e. GZ = GM sine θ (this is only true for angles of inclination up to about 15°).

$$\text{As GM} = \text{KM} - \text{KG}$$
$$\text{GM}\sin\theta = (\text{KM} - \text{KG})\sin\theta = \text{KM}\sin\theta - \text{KG}\sin\theta$$
$$\text{KM}\sin\theta = \text{KN}.$$

so GM $\sin\theta$ can be expressed as KN $-$ KG $\sin\theta$ (see KN Cross Curves on page 135).

Initial stability is the stability of the vessel in her initial position and is expressed by the metacentric height. Any reduction in GM means a loss in the ship's stability.

Dynamical stability is the measure of the work which is done when the vessel is inclined by external forces. It may be found by multiplying the vertical separation of B ang G by the displacement. For angles up to 15° it is approximately W × GM × 2 haversine θ. Another method of calculating this is shown on page 136.

STABILITY AT THE LARGER ANGLES

The metacentre can no longer be considered fixed (it is known now as the pro-metacentre). Methods other than using only the metacentric height must be employed for calculating a vessel's statical and dynamical stability. Two of these methods are:

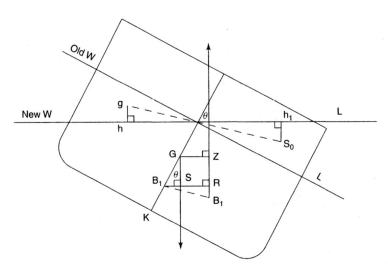

Figure 4.7

(i) use of Moseley's formula.

(ii) use of the wall-sided formula.

When a vessel of volume V heels, there is a transference of buoyancy from one side to another. In the above figure, g and g_1 are the centres of gravity of the emersed and immersed volumes or wedges of buoyancy, the volume of each being v.

The centre of buoyancy will move to B_1 (as the underwater shape has changed) and

$$BB_1 = \frac{v \times gg_1}{V}$$

BB_1 is parallel to gg_1.

This shift can be resolved into a vertical component (RB_1) and a horizontal component (BR).

$$BR - BS = GZ. \qquad BR = \frac{v \times hh_1}{V}. \qquad BS = BG \sin \theta.$$

Combining the above we get Atwood's Formula for the moment of statical stability

$$AsW \left(\frac{v \times hh_1}{V} - BG \sin \theta \right)$$

The vertical distance between G and B_1 is ZB_1 which is $RB_1 + RZ$.

$$RB_1 = \frac{v \times (gh + g_1 h_1)}{V}. \qquad RZ = BG \cos \theta.$$

The vertical distance between B and G was BG before the vessel was heeled. Multiplying the difference between the vertical distance (i.e. the vertical separation of B and G) by the displacement, we have the dynamical stability, which from the above

is $W \left(\dfrac{v \times (gh + g_1 h_1)}{V} + BG \cos\theta - BG \right)$

or $W \left(\dfrac{v \times (gh + g_1 h_1)}{V} - BG \text{ versine } \theta \right)$ this is known as Moseley's Formula

WALL SIDED STABILITY

If the ship's sides at the waterplane are parallel to one another (they are for Tankers) and the deck edge is not immersed, the wedges of immersion and emersion may be considered as being symmetrical about the ship's centre line. In such cases the 'wall sided formula' may be used to calculate the righting lever. This formula is shown on the following page and the calculation of the GZ using it is more straightforward than by Atwood's formula.

Let $b = \frac{1}{2}$ breadth of ship

θ = Angle of heel (a very small angle)

g, g_1 = centres of gravity of immersed and emersed wedges

v = volume of either of the above wedges of unit length

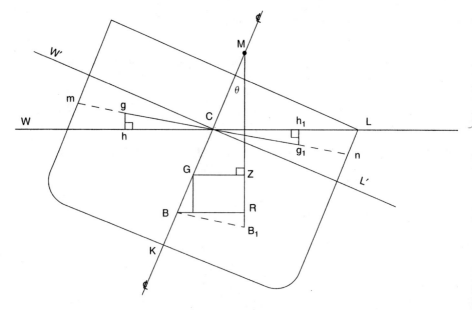

Figure 4.8

then assuming n is midpoint of LL_1

$$nL_1 = \frac{b}{2} \tan \theta$$

Now $v = \frac{1}{2} b^2 \tan \theta$

$$BR = \frac{v \times hh_1}{V}$$

$h = \frac{2}{3}$ projection of mn on line WL

$\quad = \frac{4}{3}$ of projection $CL' + L'n$ on WL

(Projection of Cn is the same as projection of $CL_1 + L_1n$ on WL)

$$hh_1 = \frac{4}{3}\left(b\cos\theta + \frac{b}{2}\tan\theta\sin\theta\right)$$

$$= \frac{4}{3}b\cos\theta\left(1 + \frac{1}{2}\tan^2\theta\right)$$

$$v \times hh_1 = \frac{2}{3}b^3 \sin\theta\left(1 + \frac{1}{2}\tan^2\theta\right)$$

$$\frac{v \times hh_1}{V} = BM\sin\theta\left(1 + \frac{1}{2}\tan^2\theta\right) = BR$$

$GZ = BR - BG\sin\theta$ and $BG = BM - GM$

$\quad = BM\sin\theta + (\frac{1}{2}BM\tan^2\theta\sin\theta) - (BM\sin\theta - GM\sin\theta)$

$GZ = \sin\theta(GM + \frac{1}{2}BM\tan^2\theta)$ known as the *'wall-sided formula'*.

WORKED EXAMPLE 26

A vessel is heeled to 20°. Assuming that she is wall-sided at the draft concerned, calculate her righting moment if the GM is 1.2 metres and the BM is 7.5 metres. The vessel's displacement is 6000 tonnes.

$$GZ = \sin\theta(GM + \frac{1}{2} BM \tan^2\theta)$$

$$= \sin 20°\left(1.2 + \frac{7.5}{2}\tan^2 20°\right)$$

$$= 0.3420(1.2 + 0.4968)$$

$$= 0.3420 \times 1.6968 \text{ metres}$$

Righting Moment $= W \times GZ$

$$= 6000 \times 0.3420 \times 1.6968$$

$$= 3482 \text{ tonnes-metres}$$

As stated at the beginning of this chapter, Ship Stability depends upon KB, KG, BM and GM. KM = KB + BM. Also KM = KG + GM.

THE EFFECT ON STABILITY CAUSED BY CHANGING THE RELATIVE POSITIONS OF B, G AND M

In all cases the vessel shown upright is in equilibrium; but only in Figs 4.9 and 4.10 is she in stable equilibrium. GM is positive. G is below M. When external forces are removed vessel will roll back to the initial upright position. The external forces could be due to wind or waves.

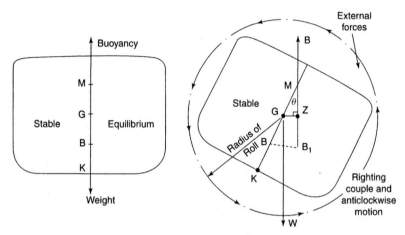

Figure 4.9 Stable equilibrium.

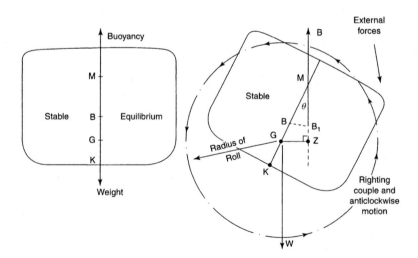

Figure 4.10 Stable equilibrium.

Transverse stability (part 1)

The conditions necessary for a vessel to be in Stable Equilibrium can be summarised as:

1. Displacement of the vessel must equal the upthrust due to buoyancy.
2. The forces of gravity and buoyancy must be on the same vertical line.
3. The centre of gravity must be below the metacentre.

Having seen that instability results when G rises above M, it may be thought that, under these circumstances, the vessel will capsize. Fortunately, this is not the case, as when the vessel starts

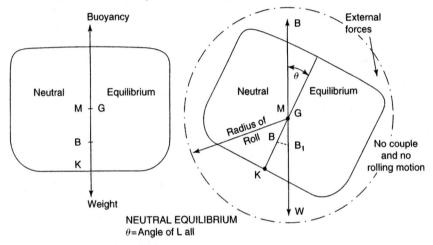

Figure 4.11 For neutral equilibrium, KG = KM. Hence GM = 0. Vessel will roll over to an angle and stay there. This angle is the Angle of Loll.

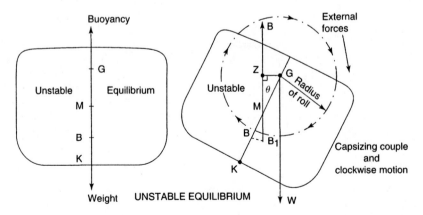

Figure 4.12 For unstable equilibrium, G is above M. GM is negative. External forces will cause the ship to roll and capsize.

to heel over, her breadth at the waterplane increases. This causes an increase in BM (BM = I/V, V remaining constant) which will eventually bring the metacentre above the centre of gravity. The angle at which the vessel comes to rest with positive stability is called the Angle of Loll. This is her new initial position and if further heeled she will return to this position. It should be noted that the metacentre is not on the centre line when the vessel is in the lolled position. Figures 4.13 & 4.14, illustrate this text along with Fig. 9.9. S/S curve.

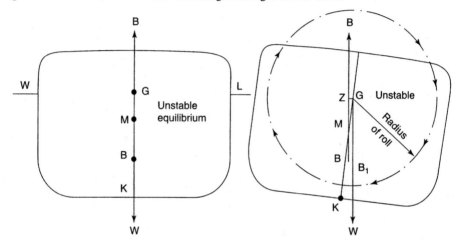

Figure 4.13

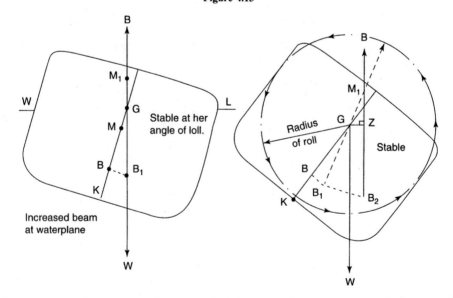

Figure 4.14

Transverse stability (part 1)

Using the 'Wall sided' formula and substituting zero for the GZ therein (i.e. the vessel is lolled) it can be shown that

$$\text{Tan angle of loll} = \sqrt{\frac{-2GM}{BM}}$$

$$\text{GM at angle of loll} = \frac{-2GM}{\cos \text{angle of loll}}$$

the GM on the right-hand side of each equation being the negative GM upright.

Vessels most prone to instability are those with deck cargoes of timber. The centre of gravity rises when fuel and water are consumed from the double bottom, and the deck cargo may absorb a considerable amount of water, if there is bad weather. Icing up of weather decks on fishing vessels also causes G to rise.

Sometimes it is necessary to reduce the angle to which the vessel is lolled, this can only be done by lowering G by either:

a) Filling a small divided double bottom tank on the low side. This will first cause G to rise due to the free surface effect of the water (see Chapter 8). As the tank becomes full G will fall and the residual list will be mainly due to unsymmetrical distribution of weight. The corresponding high side tank may now be filled.

or b) jettisoning deck cargo from the high side. Again the residual list should be mainly due to unsymmetrical distribution of weight.

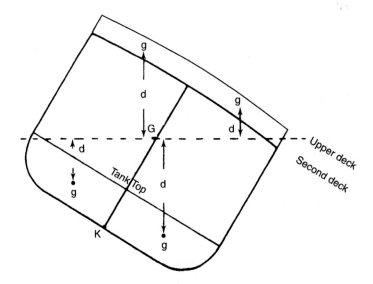

Figure 4.15

It may appear that in the above instructions, low and high should be interchanged, but Fig. 4.15 shows that the greatest VERTICAL change in the position of G is obtained by loading on the low side and discharging from the high side.

The information given on the foregoing pages provides the basic theory of transverse stability and should be thoroughly understood before passing to the practical, or calculation, aspect. Always calculate first before adding, discharging or moving weights in this situation.

The theory of taking moments was explained on page 16. Several examples of taking them in a horizontal direction have already been given. They can be taken vertically in exactly the same manner as horizontally, the following text explains this.

TAKING MOMENTS ABOUT THE KEEL

(a) Loading weights

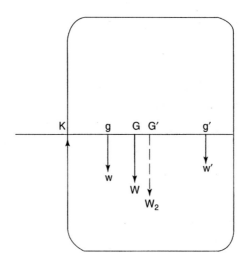

Figure 4.16

$W_2 = W + w + w'$
W is the weight of the ship

w, w' are weights loaded

$$KG' = \frac{\text{Sum of the moments about K}}{\text{Sum of the weights}} = \frac{(W \times KG) + (w \times Kg) + (w' \times Kg')}{W_2}$$

(b) Discharging weights

$$W_2 = W - w - w'$$

W is the weight of the ship

Transverse stability (part 1)

w, w' are weights discharged

$$KG' = \frac{\text{Sum of the moments about K}}{\text{Sum of the weights}} = \frac{(W \times KG) - (w \times Kg) - (w' \times Kg')}{W_2}$$

It should be noted that the moment caused by the original displacement and KG of the vessel is always considered. This forms a platform for the moment of weight calculations in the next example.

WORKED EXAMPLE 27

A vessel of 13 000 tonnes, KM 10.5 m (assumed constant), KG 9.5 m loads: 400 tonnes KG 2.9 m; 900 tonnes KG 6.0 m; 1500 tonnes KG 10.6 m; 2000 tonnes KG 8.3 m. She discharges: 700 tonnes KG 1.5 m; 300 tonnes KG 12.7 m.

Calculate the moment of statical stability if she is now heeled 8°.

Weight (tonnes)	KG (metres)	Moment (tonnes-metres)
13 000	9.5	123 500
400	2.9	1 160
900	6.0	5 400
1 500	10.6	15 900
2 000	8.3	16 600
−700	1.5	−1 050
−300	12.7	−3 810
16 800 = $\sum_1$		157 700 = $\sum_2$

KM = KG + GM.

Moment of weights table.

$\sum = $ 'summation of'

$$\text{New KG} = \frac{\sum_2}{\sum_1} = \frac{157\,700 \text{ tonnes-metres}}{16\,800 \text{ tonnes}}$$

$$KG = 9.387 \text{ metres}$$
$$KM = 10.500 \text{ metres}$$
$$\therefore GM = +1.113 \text{ metres}$$

moment of statical stability = W × GM × sin θ
= 16 800 × 1.113 × sin 8°
= *2602 tonnes-metres*

WORKED EXAMPLE 28

A vessel displacing 5800 tonnes KG 7.0 m KG 6.0 m has to load a quantity of deck cargo KG 11.0 m. What is the maximum quantity that she can load so that her GM is not less than 0.75 m.

Let w tonnes be the amount to load on deck, $\sum = $ 'summation of'

Then taking moments about the keel.

Weight	KG	Moment
5800	6.0	34 800
w	11.0	11w
$(5800 + w \text{ tonnes}) = \sum_1$		$(34\,800 + 11w) = \sum_2$

$$
\begin{aligned}
\text{KM} &= 7.00 \text{ m} \\
\text{new GM} &= 0.75 \text{ m} \\
\text{new KG} &= 6.25 \text{ m} \\
\text{because KM} &= \text{KG} + \text{GM}.
\end{aligned}
$$

$$\text{New KG} = \frac{\sum_2}{\sum_1}$$

$$6.25 = \frac{(34\,800 + 11w)}{(5800 + w)}$$

$$36\,250 + 6.25w = 34\,800 + 11w$$

$$4.75w = 1450$$

$$w = 305 \text{ tonnes}$$

TAKING MOMENTS ABOUT THE CENTRE OF GRAVITY

When dealing with horizontal movements of weight, it is usually convenient to take moments about the centre of gravity. We may also take moments about the centre of gravity when dealing with vertical movements of weight, but, unless there is only one weight involved, it is usually easier to take the moments about the keel.

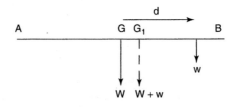

Figure 4.17

Transverse stability (part 1)

The beam AB represents a vessel of weight W tonnes acting through the centre of gravity Consider a weight of w tonnes loaded d metres from G. Take moments.

Weight	Distance	Moment
W	0	0
w	d	w × d
W + w		w × d

The distance of the new centre of gravity(G_1) from the point moments are taken (G). This is known as the shift of G or GG_1.

$$\text{This is} = \frac{\text{Sum of the moments}}{\text{Sum of the weights}}$$

i.e. $GG_1 = \dfrac{w \times d}{W + w}$

Similar expressions may be found for discharging and shifting weights, these are summarised below.

Shift of G (that is GG_1)

When loading

$GG_1 = \dfrac{w \times d}{W + w}$ where W is the vessel's displacement before loading the weight
w is the weight loaded
d is the distance of the loaded weight from the old centre of gravity

Remember, G always *moves towards* the *loaded weights*.

When discharging

$GG_1 = \dfrac{w \times d}{W - w}$ where W is the vessel's displacement before discharging the weight
w is the weight discharged
d is the distance of the discharged weight from the old centre of gravity

Remember, G always *moves away* from the *discharged weights*.

When shifting

$GG_1 = \dfrac{w \times d}{W}$ where W is the vessel's displacement (this includes the weight shifted)
w is the weight shifted
d is the distance that the weight is shifted.

Remember: G always moves in the same direction as, and parallel to, the shifted weight.

WORKED EXAMPLE 28(A)

The previous example will now be re-worked taking our moments about G

KG deck	11.0 m	KM	7.00 m
KG ship	6.0 m	KG	6.00 m
d =	5.0 m	GM now	1.00 m
		GM required	0.75 m
		GG_1 =	0.25 m upwards

Let w tonnes be the amount to load on deck

$$GG_1 = \frac{w \times d}{W + w}$$

$$0.25 = \frac{w \times 5}{5800 + w}$$

$$1450 + 0.25w = 5w$$

$$w = \frac{1450}{4.75}$$

w = *305 tonnes* as previously calculated.

Let us now consider the effect on stability when a weight is loaded off the centre line, as shown in Figs. 4.18(a) to 4.18(d).

In Fig. 4.18(a) below, the forces of gravity and buoyancy are on the same vertical line. In Fig. 4.18(b), G has moved to G_1 as the weight w has been loaded, this means that the forces of gravity and buoyancy are no longer on the same vertical line, in fact they are causing an upsetting couple. This couple forces vessel to the position shown in Fig. 4.18(c) when B and W are again on same vertical line. Vessel is now in equilibrium, at her final position,

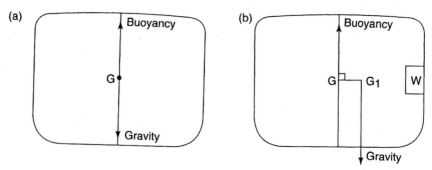

Figure 4.18

Transverse stability (part 1)

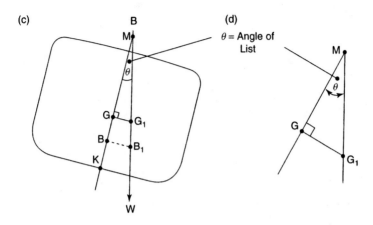

Figure 4.18 (*continued*)

Also $GG_1 = GM \tan \theta$.

When loading, discharging or shifting weights away from the centre line, the shift of G caused should always be resolved into a vertical component and a horizontal component.

WORKED EXAMPLE 29

A vessel of 6500 tonnes displacement has a KM 7.2 m and KG 6.8 m. A weight of 100 tonnes is shifted 2.3 m to port and 3.9 m upwards. If the vessel is initially upright, calculate the resulting list.

$$GG_1 = \frac{w \times d}{W}$$

for vertical shift ↑
$$GG_1 = \frac{100 \times 3.9}{6500}$$
$$= 0.06 \text{ m rise} \uparrow$$
$$KG = 6.8 \text{ m}$$
$$\overline{KG_1 = 6.86 \text{ m}}$$
$$KM = 7.20 \text{ m}$$
$$\overline{G_1M = +0.34 \text{ m}}$$

for horizontal shift →
$$GG_1 = \frac{100 \times 2.3}{6500}$$
$$= 0.0354 \text{ m to port}$$

$$\tan \theta = \frac{GG_1}{GM}$$
$$= \frac{0.0354}{0.34} = 0.1041$$
$$\theta = 5.94° \text{ to port}$$

Figure 4.19

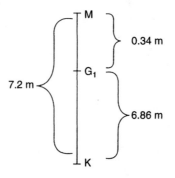

Figure 4.20

WORKED EXAMPLE 30

A vessel of 7800 tonnes displacement KM 6.8 m KG 6.0 m, measured on the centre line, is listed 4° to starboard. 400 tonnes of cargo is to be loaded into the 'tween deck KG 6.0 m. There is space 5.5 m to port and 3.0 m to starboard of the centre line. How much cargo should be loaded into each space in order that the vessel will be upright on completion?

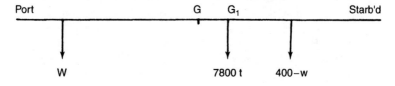

Figure 4.21

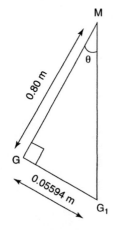

KM	6.8 m
KG	6.0 m
GM	0.8 m

$GG_1 = GM \tan \theta$

$GG_1 = 0.8 \tan 4°$

$\therefore GG_1 = 0.05594 \, m$

Figure 4.22

Let w tonnes be loaded on the *port side*, then (400 − w) tonnes will be loaded on the starboard side. For the vessel to be upright, the moments each side of the centre line will have to be equal.

i.e. $5.5w = (7800 \times 0.05594) + 3(400 - w)$

$5.5w = 436.33 + 1200 - 3w$

$w = \dfrac{1636.33}{8.5} = 192.5 \text{ tonnes.}$

i.e. 192.5 tonnes should be loaded on the port side
   207.5 tonnes should be loaded on the starboard side.

When working examples of this kind, a *sketch* will always help the student to see what has to be done. Weights loaded should always be indicated by arrows pointing downwards and weights discharged by arrows pointing upwards.

It is also useful to erect a perpendicular line about four or five centimetres in height and place on it KB, KG, KM etc. As these values are calculated, insert them onto this perpendicular line. See Figure 4.20 as an example of this.

CHAPTER FOUR

Transverse Stability (Part 2)

PRINCIPLE OF SUSPENDED WEIGHTS

When a weight is suspended by a ship's derrick, its centre of gravity is to be considered as being *at the derrick head*, and it will remain at the derrick head as long as the weight is suspended. In other words it does not matter if the weight is 1 cm or 10 metres above the deck, its C of G is still be be considered at the derrick head.

WORKED EXAMPLE 31

A vessel of 9920 tonnes displacement, KM 7.8 m is to load two 40 tonne lifts on deck KG 13.5 m and 5.5 m each side of the centre line, by means of her heavy lift derrick whose head is 21.0 m above the keel and maximum swing out 15 m from the centre line. What should be the vessel's maximum KG *before loading*, if the list during loading is not to exceed 5°? (The inboard weight is to be loaded first.)

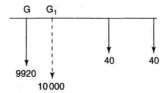

Figure 4.23

To find the horizontal shift of G.

Take moments about the centre line when the maximum list occurs (See Figure 4.24).

Weight	Distance	Moment
9920	0	0
40	5.5	220
40	15.0	600
$10\,000 = \sum_1$		$820 = \sum_2$

$\sum$ = 'summation of'

$$GG_1 = \frac{\sum_2}{\sum_1} = \frac{820}{10\,000} = 0.082 \text{ m}$$

$$GM \tan \theta = GG_1$$

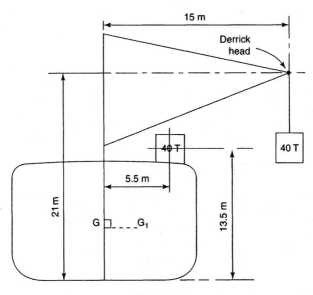

Figure 4.24

$$GM \tan 5° = 0.082$$
$$\therefore GM = 0.9373 \text{ m}$$
$$KM = 7.800 \text{ m}$$
$$KG = KM - GM$$
$$= 7.800 - 0.9373$$
$$\therefore KG = 6.8627 \text{ m}$$

Taking moments about the keel............Let the original KG be at x metres.

Weight	KG	Moment
9920	x	9920x
40	13.5	540
40	21.0	840
$10\,000 = \sum_3$		$(9920x + 1380) = \sum_4$

Final $KG = \dfrac{\sum_4}{\sum_3}$

$$6.8627 = \frac{(9920x + 1380)}{10\,000} \qquad \text{Hence } x = 6.779 \text{ metres, say } 6.78 \text{ m}$$

In the examples considered so far the position of the ship's centre of gravity has always been given. How was this originally found? It could have been calculated by knowing the weights of all the steel plating, the wood and outfit weight and the machinery weight and then taking moments about the base. This would have been laborious and time taking.

In practice the position of the ship's centre of gravity is found via an Inclining Experiment conducted on the vessel, just a few days before being completed by the Shipbuilders.

INCLINING EXPERIMENT OR STABILITY TEST

This is carried out to find the Lightweight (ship's weight when empty), the VCG at this Lightweight and the LCG about amidships at this Lightweight. At the time of the experiment the main aim is to estimate the GM for the ship as inclined. Recall that VCG is same as KG.

Hydrostatic information will give the displacement and KM for the ship as inclined. The difference KM and obtained GM then gives the KG for the ship as inclined. Hydrostatics will also give the LCB from amidships for the ship as inclined. Using trim calculations will then give the ship's LCG as inclined.

Adjustments now have to be made because the ship is not fully completed. Weights have to go on board. Weights have to come off the ship. After a moment of weight calculation the final values for the Lightweight, its VCG and its LCG are obtained.

On the Test a plumbline is suspended from a hatch coaming on the centreline down to a lower hold where a graduated batten is set up horizontally. A known weight 5 t to 15 t (depending on size of ship) is then shifted transversely across the Upper Deck. See Fig. 4.25. This causes ship to list and plumbline to move across the batten. The deflection is measured. See Figs 4.26 and 4.27.

To obtain good results two pendulums are used, one for'd and one aft. Also four weights can be used to give eight readings on the two battens. A mean deflection is then used in the GM calculation. See Worked Example 32.

To ensure that the Test results are reliable:

    1. The ship must be upright at the start of the Test.

    2. There should be little or no wind.

    3. The moorings should be slack and the vessel well clear of the jetty or drydock side.

    4. Free surface of liquids should be kept to an absolute minimum. Slack tanks if possible should be emptied or completely filled.

    5. A note of 'Weights on' and 'Weights off' each with a VCG and LCG must be made.

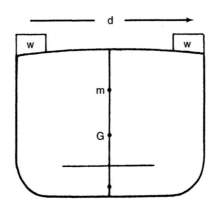

Figure 4.25

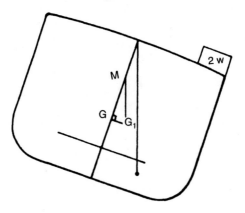

Figure 4.26

6. Only those directly concerned with the with the Test should be on board. If the weights are lifted manually across the Deck then those doing the shifting should stand on the centreline as the pendulum readings are made.

7. The density of the water in which the vessel is floating should be measured for'd, amidships and aft. The mean density is used to estimate the ship's displacement W in tonnes.

$$\frac{GM}{GG_1} = \frac{\text{length of the plumbline}}{\text{deflection of the plumbline}} \quad \text{(Similar triangles)}$$

$$\text{But } GG_1 = \frac{w \times d}{W}$$

$$\text{So } GM = \frac{w \times d \times \text{length of the plumbline}}{W \times \text{deflection of the plumbline}}. \quad \text{So } GM = \frac{w \times d}{W \tan \theta}$$

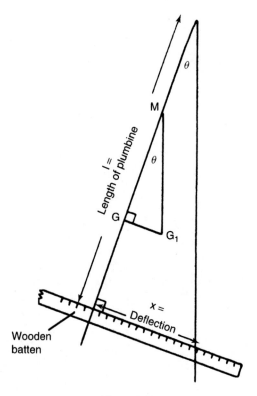

Figure 4.27

WORKED EXAMPLE 32

When a vessel of 5300 tonnes displacement KM 7.7 m is inclined by shifting 10 tonnes 16 m, it is noted that the mean deflection of a plumbline 12 m long is 33.25 cm. What is her KG and inclined angle θ?

$$GM = \frac{w \times d}{W \tan \theta}$$

$$\therefore GM = \frac{10 \times 16 \times 12 \times 100}{5300 \times 33.25}$$

$$\therefore GM = 1.09 \, m$$

$$KM = 7.70 \, m$$

$$\overline{KG = 6.61 \, m}$$

$$\tan \theta = \frac{x}{l} = \frac{33.25}{1200} = 0.0277 \quad \therefore \theta = 1.59°$$

Transverse stability (part 2) 79

MAXIMUM PERMISSIBLE DEADWEIGHT–MOMENT CURVE

This is simply a graph of displacements against maximum-deadweight moments. Figure 4.28 shows a typical curve with the ship's draft superimposed onto the vertical axis. The deadweight-moment has been calculated about the keel. Note that this ship's Lightweight is 370 tonnes.

It is very important to realise that the total deadweight–moment at any displacement must not under any circumstances exceed the maximum deadweight–moment at this displacement. In other words the x,y intercept must NOT occur in the shaded area of Fig. 4.28. If it does, then deficient stability will occur for this ship.

For example when the displacement is 1000 tonnes, then a deadweight–moment of 1200 tm is acceptable. However, a deadweight-moment of 1480 tm would mean this ship has deficient stability and is not acceptable. A deadweight–moment value of 1375 tm would be just allowable.

If the maximum KG for this 1375 tm is required it is simply:

$$\text{max KG for the dwt} = \frac{\text{Deadweight–Moment}}{\text{Deadweight}} = \frac{1375}{(1000 - 370)}$$
$$\uparrow$$
$$\therefore \text{dwt KG} = 2.18 \text{ m above base.} \quad \text{(displacement-lightweight)}$$

Worked Example 33 shows the calculations involved and the use of the Deadweight–Moment curve.

WORKED EXAMPLE 33

USE OF DWT–MOMENT CURVE

Use Fig. 4.28: Simplified Stability - Deadweight Moment Curve. (Ensure that all points plotted on the diagram can be identified in the working.)

The load displacement is 1175 tonnes.

The ship's present condition: displacement 800 tonnes.
                                 Deadweight moment 600 tm.
Cargo to be loaded: 250 tonnes at Kg 2.8 m.

From the above condition:

(a) Determine the maximum weight of cargo that can be taken at Kg 4.5 m so that stability is adequate.

(b) It is anticipated that 30 tonnes of fuel and water at Kg 1.6 m will be used on passage producing a free surface moment of 100 tm.

Determine the maximum weight of cargo that can be loaded at Kg 4.5 m so that stability is adequate on arrival at the discharge port.

(a)

Item	Weight	Kg	Deadweight Moment
Present Condition (as given)	800	–	600......see Point 1 on Fig. 4.28
Cargo	w	4.5	4.5w
Cargo	250	2.8	700
	(1050 + w)		(1300 + 4.5w)

Let $w = 125$ t: The $W = 1050 + 125 =$ the given load displacement of 1175 t. Let $w = 125$ t
∴ $W = 1175$ t and Dwt-mmt $= 1863$ t · m.

This plots okay as Point 2 on the Dwt-mom't curve on Fig. 4.28.

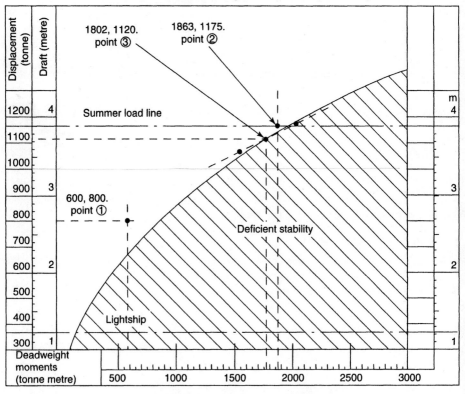

Summer load displacement 1175 tonne,
Lightship displacement 370 tonne.

Figure 4.28

Transverse stability (part 2)

Consequently, w of 125 t @ kg of 4.5 m is adequate.

(b)

Item	Weight	Kg	Deadweight Moment
Present Condition (as given)	800 t		600
Cargo	250 t	2.8	700
fuel & water	−30 t	1.6	−48
free surface effects	−	−	+100
Cargo	+w	4.5	4.5w
	(1020 + w)		(1352 + 4.5w)

Substitute for weight w:

Let w = 50 t this results in 1070 t against 1577 t.m. positive stability

Let w = 150 t this results in 1170 t against 2027 t.m. deficient stability

Let w = 100 t this results in 1120 against 1802 t.m. intersection region. (see point ③ on Figure 4.28)

Plot these co-ordinates onto Fig. 4.28. Observe intersection with the curve and loci of points.

Summary: 100 t of cargo can only be loaded at Kg of 4.5 m because of free surface moments of fuel & water together with raising of G due to using these liquids on voyage.

ROLLING

The reader will now be aware that the stability of the vessel depends on the righting moment which can be varied either by changing the GZ or the displacement. A change in either of these can also affect the rolling period of the ship and some notes on this follow.

Wave period is the time in seconds between successive crests, or troughs, passing a fixed point.

Apparent wave period is the time in seconds between successive crests or troughs, passing an observer on board ship.

Ship's period is the time in seconds taken by a vessel doing a complete roll (e.g. Port-Starboard - Port). This is denoted by T_R.

A dangerous state of affairs can arise if the ship's still water period and the apparent wave period are equal. This is known as Synchronism and, if allowed to continue, it could capsize a vessel.

It should be noted that the movement of a vessel when rolling is similar to that of a pendulum, this being so, an expression can be formed for the vessel's period when in still water as follows:

Still water period $T_R = 2\pi\sqrt{\dfrac{k^2}{GM \cdot g}}$ seconds.

$\dfrac{\pi}{\sqrt{g}} \simeq 1$. So $T_R = 2\sqrt{\dfrac{k^2}{GM}}$ secs approx.   Hence $T_R \propto \dfrac{1}{\sqrt{GM}}$ approx.

where k is the radius of gyration. This can be increased by stowing weights well away from the ship's centre of gravity, g is acceleration due to gravity, GM in metres.

k can be estimated as being $0.35 \times$ Breadth Moulded.

It can be seen that if the radius of gyration remains constant, the period can be altered by increasing or decreasing the GM. As the ship's period will normally be greater than the wave period, an increase in GM will bring the ship's period nearer to the wave period and possible synchronism. It would seem that a small GM is safer than a large GM Certainly, a vessel's movement is easier and more comfortable when the *GM is small*, but she is *'tender'* and could possibly be made unstable should cargo or ballast shift. A vessel with a *large GM* is *'stiff'* and has an uncomfortable, jerky movement in a seaway. However, should cargo or ballast be liable to shift, a large GM is safer than a small GM.

Examples for T_R:

'tender ship' $T_R = 30$ to 35 seconds.

'stiff ship' $T_R = 8$ to 10 seconds.

comfortable ship $T_R = 20$ to 25 seconds.

What is a reasonable GM? Below are shown typical GM values for several ship-types when fully-loaded.

Ship-type	GM when fully-loaded
General cargo ships	0.30 m to 0.50 m.
Oil Tankers to VLCCS	0.30 m to 1.00 m.
Containers ships	1.5 m approx.
Ro-Ro vessels	1.5 m approx.
Bulk-ore carriers	2 m to 3 m.
All ship-types	Minimum GM = 0.15 m.

When loading, weights should be 'winged out' and not concentrated on the centreline. This also affects the ship's period, but not to the same extent as does the GM.

If it is found that a vessel is tending to synchronize, then the apparent wave period should be changed either by altering course or altering speed.

Transverse stability (part 2)

The onset of synchronism can be recognized by the increase in the angle to which the vessel is rolling. At each oscillation (half roll) this angle increases by about $1\frac{1}{2}$ times the wave slope. It will be seen that this angle could soon reach dangerous limits in synchronous conditions.

The operation of stabilizers either gyro type or flume tank is beyond the present remit of this book. Hydraulic-fin stabilisers also reduce rolling of ships.

WORKED EXAMPLE 33A

A ship has the following information: Displacement is 8775 tonnes, natural rolling period TR is 17.5 secs, GM is 1.12 m.

(a) Determine the new natural rolling period after:

  1750 tonnes are added 3.85 m above ship's KG with

  450 tonnes are discharged 2.75 m below ship's KG.

(b) What approximately is the Breadth moulded for the ship? Assume KM is constant before & after changes of loading.

(a) $T_{R_{(1)}} = 2\pi \sqrt{\dfrac{k_{(1)}^2}{g \cdot GM_{(1)}}}$ So $k_1^2 = \left(\dfrac{T_{R_{(1)}}}{2\pi}\right)^2 \times g \times GM_{(1)}$

$\therefore k_1^2 = \left(\dfrac{17.5}{2 \times \pi}\right)^2 \times 9.81 \times 1.12$ $k_1^2 = 85.21\,m^2$. $k_1 = 9.23\,m$.

$I_1 = W \cdot k_1^2 = 8775 \times 85.21 = 747\,718$ tonnes $\cdot m^2$.

Weight	Lever (KG)	Moment of Weight	Lever (KG)^2	Moment of Inertia
8775	0	0	0	747 718
+1750	+3.85	+6738	+14.82	+25 935
−450	−2.75	−1238	+7.56	−3 402
10 075	+0.55	+5500		770 251 t.m^2

$-W_2 \times y^2 = 10\,075 \times 0.55^2 = -3\,048\,t.m^2$

New Inertia $= I_2 = 767\,203\,t.m^2$

KM is constant

G has risen by 0.55 m, so *new GM* $= 1.12 - 0.55 = 0.57\,m = GM_2$

$k_2^2 = \dfrac{I_2}{W_2}$ $\therefore k_2^2 = \dfrac{767\,203}{10\,075} = 76.15\,m^2$. $\therefore k_2 = 8.73\,m$

$$T_{R_{(2)}} = 2\pi \sqrt{\frac{k_2^2}{g \cdot GM_2}} = 2\pi \sqrt{\frac{76.15}{9.81 \times 0.57}} = 23.19 \text{ seconds. (i.e. reasonably comfortable rolling period).}$$

(b) $k = 0.35 \times$ Br.mld approx. average $k = \dfrac{k_1 + k_2}{2} = 8.98$ m.

$\therefore$ Approx Br.mld. $= \dfrac{8.98}{0.35} = 25.7$ m.

LOSS OF UNDERKEEL CLEARANCE (UKC) CAUSED BY A *STATIC* VESSEL HEELING OR LISTING

To understand the implication of this, DRAFT must first be defined. It is the minimum depth of water that is necessary to float the ship, which is the distance from the waterline to the lowest point of the ship.

In the diagram, where for clarity only half of the ship is shown, it is assumed that the vessel will heel about the keel. This is not quite correct, but it gives an answer sufficiently close to the

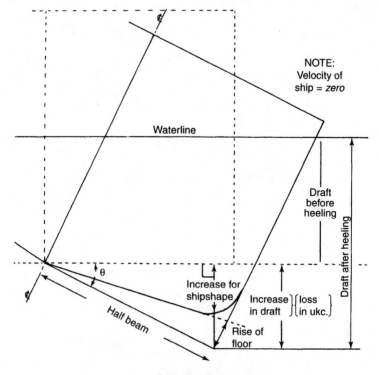

Figure 4.29

true answer for practical purposes. If the vessel is boxshaped the increase in draft is 1/2 beam sin θ. This $\frac{1}{2}$. b. sin θ is of course a loss in underkeel clearance.

When dealing with a shipshape, a rise of floor may have to be considered. Although this is measured along the ship's side, for practical purposes the increase in the draft is assumed to be decreased by the rise of floor, this gives a slight safety margin. Nowadays rise of floor is no longer fitted. It is almost obsolete.

WORKED EXAMPLE 34

Draw graphs of Loss of ukc against angle of heel of for θ up to 8° for the following stationary vessels:

(a) General cargo ship of 20 m-beam.
(b) Panamax vessel of 32.25 m-beam.   } Assume rise of floor is zero.
(c) ULCC having a beam of 70 m.
(d) When upright, if the ukc was 1.25 m, then evaluate the angle of heel of at which each of the above vessels would go aground in way of the bilge plating.

Loss of ukc = $\frac{1}{2} \times b \times \sin\theta$.   Substitute for b and θ.

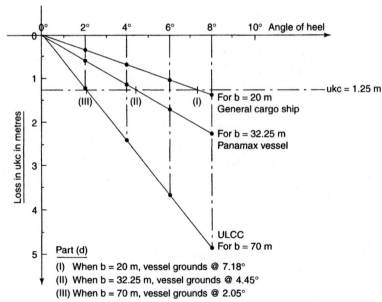

Figure 4.30

Angle of heel	0°	2°	4°	6°	8°
for b = 20 m	0	0.35	0.70	1.05	1.39 mtrs.
for b = 32.25 m	0	0.56	1.12	1.69	2.24 mtrs.
for b = 70 m	0	1.22	2.44	3.66	4.87 mtrs.

WORKED EXAMPLE 35

A VLCC has a C_B value of 0.830. Her ratio of water depth to static even keel draft is 1.10. State her maximum squats at the bow when:

(a) she is operating in a confined channel such as a river.

(b) she is operating in open water conditions. Use a speed range of 0 to 12 knots.

(c) Draw graphs of max.squat α ship speed for these two situations.

(d) If static ukc was 0.75 m at what speeds would this VLCC have grounded?

Open water: $`\delta`_{max} = \dfrac{C_B \times V_K^2}{100}$ metres

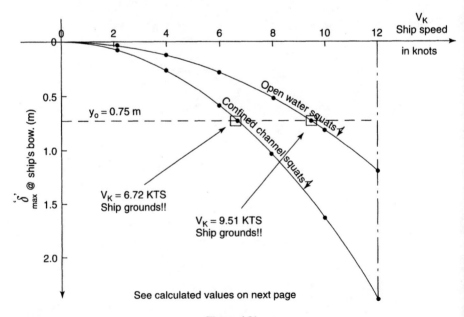

Figure 4.31

Transverse stability (part 2)

V_K (kts)	0	2	4	6	8	10	12
'δ'$_{max}$ (m)	0	0.03	0.13	0.30	0.53	0.83	1.20

Confined Channel: $\delta_{max} = \dfrac{C_B \times V_K^2}{50}$ metres.

V_K (kts)	0	2	4	6	8	10	12
'δ'$_{max}$ (m)	0	0.07	0.27	0.60	1.06	1.66	2.39

ANGLE OF HEEL WHEN A SHIP TURNS

An Angle of Heel occurs when a ship turns. This is because forces act an the rudder and because centrifugal forces and moments develop.

If a ship's rudder is put hard over to PORT, then the ship will, due to forces on the rudder, heel over slighty to PORT, say θ_1.

However, due to the centrifugal forces involved, the ship will heel to STARB^D, say θ_2. The angle θ_2 will always be greater then θ_1. Consequently with PORT rudder helm, the final angle of heel whilst turning the ship will be to STARB^D and vice versa.

WORKED EXAMPLE 36

A ship has the following particulars:

Displacement W = 5000 t, rudder area $A_R = 12\,m^2$, VCB to rudder cg = 1.6 m (NL), GM = 0.24 m, rudder helm 35° to PORT, ship speed = 16 kts. Calculate the angle of heel whilst the ship turns at the given speed, due to only forces on the rudder. Assume 1 kt = 1852 m.

$$V = 16 \times \frac{1852}{3600} = 8.231 \text{ m/sec.} \quad \text{Let } \alpha = \text{angle of rudder helm.}$$

$$F = 580 \times A_R \times V^2 \text{ Newtons.} \qquad F_t = F \cdot \sin\alpha \cdot \cos\alpha$$

$$F_t = 580 \times A_R \times V^2 \times \sin\alpha \cdot \cos\alpha = 580 \times 12 \times 8.231^2 \times \sin 35° \cos 35°$$

$\therefore F_t = 580 \times 12 \times 67.75 \times 0.5736 \times 0.8192 \quad \text{Let } \theta = \text{angle of heel.}$

$\therefore F_t = 221,560\,N \text{ or } 221.6\,kN$

$$\tan\theta = \frac{F_t \times NL}{W \times g \times GM} = \frac{221.6 \times 1.60}{5000 \times 9.81 \times 0.24}$$

$\therefore \tan\theta = 0.03011. \qquad \text{Thus } \theta = 1.78° \text{ to port}$

WORKED EXAMPLE 37

A vessel turns to PORT in a circle of diameter of 200 m at a speed of 16 kts. VCB to VCG = 0.88 m, $GM_T = 0.78$ m, $g = 9.81$ m/sec². Assume 1 kt = 1852 m.

If the forces on the rudder cause an angle of heel to PORT of 1.05°, then proceed to calculate the final angle of heel at the given speed when centrifugal forces are also considered.

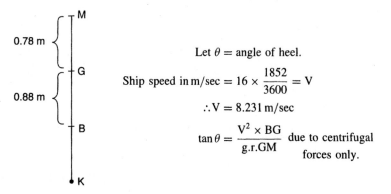

Let θ = angle of heel.

Ship speed in m/sec = $16 \times \dfrac{1852}{3600} = V$

$\therefore V = 8.231$ m/sec

$\tan \theta = \dfrac{V^2 \times BG}{g.r.GM}$ due to centrifugal forces only.

Figure 4.32

$$\therefore \tan \theta = \dfrac{8.231^2 \times 0.88}{9.81 \times \dfrac{200}{2} \times 0.78} = 0.07791.$$

$\therefore \theta = 4.45°$ to STARB^D due to centrifugal forces only.

Final $\theta = 4.45°$(stb) $- 1.05°$(Port)

$\qquad = 3.40°$ to STARB^D

SUMMARY

With angle of heel whilst turning

1. Two sets of forces are involved, namely forces on the rudder and centrifugal forces.

2. If rudder helm is set to PORT, then final angle of heel will be to STARB^D, and vice versa.

3. If KG is reduced by shifts in loading, then BG will decrease and GM increases. Consequently, $\tan \theta$ will decrease, and so angle of heel whilst turning will be LESS.

CHAPTER FIVE
Longitudinal Stability, i.e. Trim

Trim is the difference between the draft at the for'd perpendicular (FP) and the draft at the aft perpendicular (AP). If there is no difference, ship is on even keel.

Change of Trim is the difference between the original trim and the final trim.

Centre of Flotation (LCF) or Tipping centre (TC) is the geometrical 2-dimensional centre of the waterplane. It is the point about which the ship trims. In effect it is the fulcrum of the waterplane.

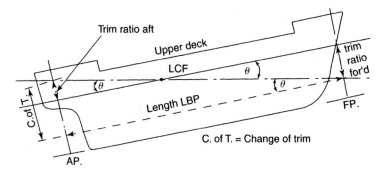

Figure 5.1

In the above diagram θ is the angle of trim Fig. 5.1 indicates that:

$$\text{Tan}\,\theta = \frac{\text{Change of Trim}}{\text{LBP}}$$

$$\text{also Tan}\,\theta = \frac{\text{Trim ratio for'd}}{\text{LCF to FP}}$$

$$\text{also Tan}\,\theta = \frac{\text{Trim ratio aft}}{\text{LCF to AP}}$$

By using Fig. 5.1 and transposing the above formulae

$$\text{Change of Trim} \times \frac{\text{LCF to FP}}{\text{LBP}} = \textit{trim ratio for'd in metres.}$$

$$\text{and Change of Trim} \times \frac{\text{LCF to AP}}{\text{LBP}} = \textit{trim ratio aft in metres.}$$

To find the moment to change trim one centimetre (MCT 1 cm) which is the moment required to change the trim by one centimetre. This is a very important value with trim calculations. It is used to calculate the change of Trim.

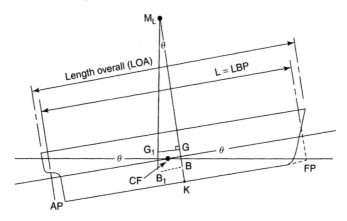

Figure 5.2

In the figure above

M_L is the longitudinal metacentre.

G is centre of gravity before trimming.

G_1 is centre of gravity after trimming.

B is centre of buoyancy before trimming.

B_1 is centre of buoyancy after trimming.

θ is the angle of trim.

Note:

$\dfrac{LBP}{LOA} = 96\%$ approximately. Amidships is midway between AP and FP i.e. LBP/2. It is *not* the midway or midlength of LOA on most merchant ships.

Now
$$\frac{GG_1}{GM_L} = \tan\theta = \frac{\text{Change of trim}}{\text{Length of the vessel}}$$

$$GG_1 = \frac{GM_L \times \text{change of trim}}{\text{Length of vessel}}$$

$$GG_1 = \frac{GM_L}{100L} \quad \left(\text{if change of trim is 1 cm: 1 cm} = \frac{1}{100}\text{ m}\right)$$

or
$$\frac{w \times d}{W} = \frac{GM_L}{100L} \quad \left(\text{as } GG_1 = \frac{w \times d}{W} \text{ and } w \times d \text{ is moment in tonnes-metres} \atop W \text{ is displacement in tonnes}\right)$$

or
$$w \times d = \frac{W \times GM_L}{100L}$$

Longitudinal stability, i.e. Trim

but w × d is the moment which has changed the trim one centimetre.

$$\therefore \text{MCT 1 cm} = \frac{W \times GM_L}{100\,L} \text{ tonnes metres (tm)}.$$

Approximation of MCT 1 cm for a box-shaped vessel

MCT 1 cm $= \dfrac{W \times GM_L}{100 \times L}$.   Now $GM_L \simeq BM_L$ and $BM_L = \dfrac{L^2}{12 \cdot d}$

Hence MCT 1 cm $\simeq \dfrac{(1.025 \times L \times B \times d)}{100 \times L} \times \left(\dfrac{L^2}{12d}\right) = \dfrac{1.025 \cdot L \cdot B}{100} \times \dfrac{L}{12}$

$\therefore$ MCT 1 cm $\simeq \dfrac{1.025 \cdot T}{1.025} \times L/12 = T \times \dfrac{100 \cdot T}{12 \times 1.025 \cdot B}$ where T = TPC.

Thus, MCT 1 cm for a box-vessel $\simeq \dfrac{100 \cdot T^2}{12.3B} = \dfrac{8 \cdot 1 \times T^2}{B} \simeq \dfrac{8 \times T^2}{B}$ tm/cm.

For *Oil Tankers*, a figure of MCT 1 cm $= \dfrac{7.8 \times T^2}{B}$ is a good approximation.

For *General Cargo ships*, MCT 1 cm $= \dfrac{7.2 \times T^2}{B}$ approximately.

WORKED EXAMPLE 38

For a General Cargo Ship, the TPC is 21.05 and the Breadth moulded is 19.75 m. Calculate the approximate MCT 1 cm.

$$\text{MCT 1 cm} = \frac{7.2 \times (TPC)^2}{B} = \frac{7.2 \times 21.05^2}{19.75}$$

$\therefore$ MCT 1 cm = 161.54 tm/cm approximately.

Procedure for Trim problems

1. Make a *sketch* from the given information. Label given values.

2. Estimate the mean bodily sinkage, remembering to convert the units to metres.

3. Calculate the Change of Trim, using each lever measured from the LCF position. Convert units to metres.

4. Calculate the trim ratio forward and aft about the LCF position, measuring to the AP and the FP.

5. Collect all these values together to estimate the final end drafts as follows:

each Final end draft = original draft + mean bodily sinkage/rise ± trim ratio.

6. In practice, it is usual to round off these final end drafts to *two* decimal figures.

WORKED EXAMPLE 39

A weight of 54 tonnes is shifted from No. 1 to No. 2 hold, a distance of 20 metres, on a vessel MCT 1 cm 120 tonnes-metres. Calculate the change of trim.

$$\text{Change of trim} = \frac{\text{Moment being caused}}{\text{MCT 1 cm}}$$

$$= \frac{54 \times 20}{120}$$

∴ Change of trim = 9 cm by the stern.

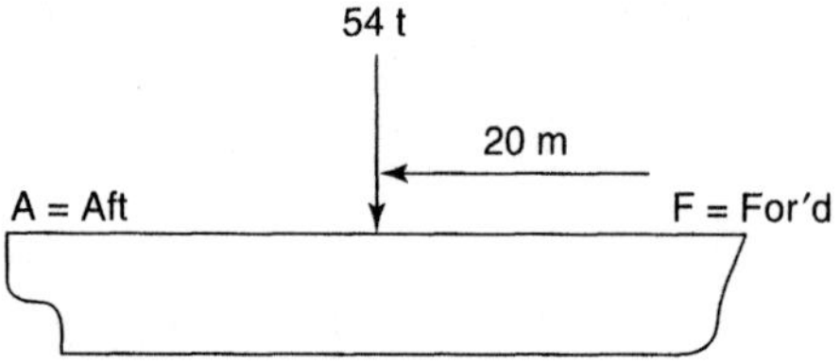

Figure 5.3

N.B. The positions of Nos 1 and 2 holds relative to the centre of flotation are immaterial. The *distance* and *direction* the weight is shifted are the important factors.

WORKED EXAMPLE 40

A vessel 120 m long MCT 1 cm 100 tonnes-metres, TPC 25 is drawing 6.00 m F 6.60 m A. A weight of 250 tonnes is loaded 12 m forward of the centre of flotation which is 2 m abaft amidships. Calculate the new end drafts forward and aft.

$$\text{Bodily sinkage} = \frac{\text{Weight loaded}}{\text{TPC}} = \frac{250}{25} = 10 \text{ cm} = 0.10 \text{ m}$$

$$\text{Change of trim} = \frac{\text{moment caused}}{\text{MCT 1 cm}} = \frac{250 \times 12}{100} = 30 \text{ cm by the head.}$$

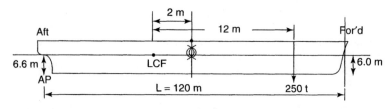

Figure 5.4

Longitudinal stability, i.e. Trim

Trim ratio forward due to change of trim = $\dfrac{62}{120} \times 30 = 15.5$ cm increase.

Trim ratio aft due to change of trim = $\dfrac{58}{120} \times 30 = 14.5$ cm decrease.

	A	F
Original draft	6.600	6.000
Sinkage	0.100	0.100
	6.700	6.100
Trim ratio	−0.145	+0.155
Final end drafts	6.555 m	6.255 m
	say 6.56 m	say 6.26 m.

WORKED EXAMPLE 41

A vessel of 6600 tonnes displacement 120 m in length, GM_L 140 m is drawing 4.8 m F 4.5 m A. The centre of flotation is 2 m abaft amidships. How much cargo should be discharged from No. 2, Hold which is 16 m forward of amidships, so that the vessel would be trimmed 15 cm by the stern?

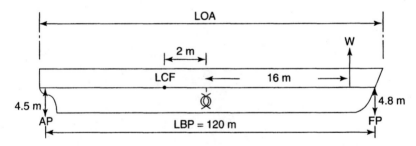

Figure 5.5

Let w tonnes be the weight to discharge from No. 2. Hold this will cause a moment of 18 w tonnes-metres about the centre of flotation.

Present draft	4.80 m F	$\text{MCT 1 cm} = \dfrac{W \times GM_L}{100\,L}$
	4.50 m A	
Present trim	0.30 m by the head ↻	$= \dfrac{6600 \times 140}{100 \times 120}$
Required trim	0.15 m by the head ↻	
Change of trim required	0.45 m by the stern ↻	= 77 tonnes-metres

The moment required to cause the above change of trim is 45 × 77 tonnes-metres.

$$\text{The moment caused} = \text{The moment required}$$
$$18 \, w = 45 \times 77$$
$$w = 192.50 \text{ tonnes}$$

WORKED EXAMPLE 42

A vessel 150 m in length, 18 m in breadth, MCT 1 cm 150 tonnes-metres, TPC 25 is drawing 6.35 m F 6.65 m A and loads the following:

230 tonnes in No. 1 hold	50 m forward of the centre of flotation
800 tonnes in No. 3 hold	20 m forward of the centre of flotation
500 tonnes in No. 4 hold	21 m abaft of the centre of flotation
She discharges 200 tonnes from No. 2 hold	36 m forward of the centre of flotation
She discharges 105 tonnes from F.P tank	60 m forward of the centre of flotation

The centre of flotation is 5 m abaft amidships. Calculate the new end drafts.

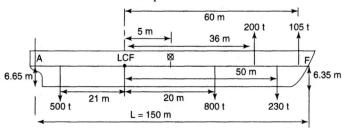

Figure 5.6

	Weight W	Distance from LCF	Moment in tonnes-metres	
			Aft	Forward
	230	50 F	–	11 500
	800	20 F	–	16 000
	500	21 A	10 500	–
	−200	36 F	7 200	–
	−105	60 F	6 300	–
Net weight loaded	1225 tonnes		24 000	27 500
				24 000
		Resultant Moment, tm Forward =		3 500

Longitudinal stability, i.e. Trim

$$\text{Mean bodily sinkage} = \frac{\text{Weight loaded}}{\text{TPC}}$$

$$= \frac{1225}{25}$$

$$= 49 \text{ cm} = 0.49 \text{ m}. \downarrow$$

$$\text{Change of trim} = \frac{\text{Resultant moment}}{\text{MCT 1 cm}}$$

$$= \frac{3500}{150}$$

$$= 23.33 \text{ cm by the head} \circlearrowright$$

$$\text{Trim ratio forward due to change of trim} = \frac{80}{150} \times 23.33$$

$$= 12.44 \text{ cm increase.}$$

$$\text{Trim ratio aft due to change of trim} = \frac{70}{150} \times 23.33$$

$$= 10.89 \text{ cm decrease.}$$

	A	F
Old drafts	6.650 m	6.350 m
Sinkage	0.490 m	0.490 m
	7.140 m	6.840 m
Trim ratio	−0.109 m	+0.124 m
New end drafts	7.031 m	6.964 m
	say 7.03 m	say 6.96 m

WORKED EXAMPLE 43

A vessel of 9000 tonnes displacement, length 120 m, GM_L 160 m is trimmed 8 cm by the head. After loading 300 tonnes 24 metres forward of the amidships, 400 tonnes 30 metres abaft amidships and discharging 200 tonnes from amidships, it is noted that she is trimmed 18 cm by the stern.

Calculate the position of the centre of flotation.

Trim by the head is the same as trim by the Bow.

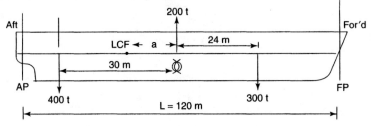

Figure 5.7

$$\text{MCT 1 cm} = \frac{W \times GM_L}{100L}$$

$$= \frac{9000 \times 160}{100 \times 120}$$

$$= 120 \text{ tonnes-metres}$$

Original trim = 8 cm by the head ↻

Final trim = 18 cm by the stern ↻

Change of trim = 26 cm by the stern ↻

Let us assume that the centre of flotation is 'a' metres abaft amidships. Then taking moments about the centre of flotation

The moment caused by loading and discharging = The moment required to change the trim 26 cm.

$$400(30 - a) + 200a - 300(24 + a) = 120 \times 26$$

$$12\,000 - 400a + 200a - 7200 - 300a = 3120$$

$$500a = 1680$$

$$a = 3.36 \text{ metres}$$

Thus the LCF is 3.36 m aft of amidships.

If the assumption, that the LCF was abaft amidships, had been incorrect, the result would have been a negative quantity. This would indicate that the centre of flotation was the opposite side of amidships to that which it had been assumed.

TO FIND WHERE TO PLACE A WEIGHT TO KEEP THE DRAFT CONSTANT AT ONE OF THE PERPENDICULARS.

Assuming that the draft aft is to be kept constant.

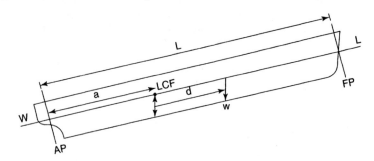

Figure 5.8

Then in Figure 5.8:

w is the weight loaded in tonnes.

Longitudinal stability, i.e. Trim

d is the distance in metres from the centre of flotation that the weight is to be loaded.

a is the distance in metres of the after perpendicular from the centre of flotation.

L is the LBP of the vessel in metres.

If w is loaded at the centre of flotation, the loading sinkage will be $\frac{w}{TPC}$. For the draft aft to remain constant, a change of draft equal to the sinkage will have to be caused, this will be caused by shifting w tonnes d metres.

$$\text{Trim ratio aft} = \text{mean bodily sinkage}$$

$$\text{Then} \quad \frac{w \times d \times a}{MCT\ 1\ cm \times L} = \frac{w}{TPC}$$

$$\text{Hence } d = \frac{MCT\ 1\ cm \times L}{TPC \times a}$$

Note: This formula will only hold good if the hydrostatic data does not change. This is acceptable when weight 'w' is small.

WORKED EXAMPLE 44

A vessel 150 m in length, MCT 1 cm 140 tonnes-metres TPC 20, centre of flotation 5 m abaft amidships, is loading at a certain port. It is noted that she reaches her required draft aft when there are still several tonnes of cargo on the quay.

Where should the cargo be loaded in one block so as to maintain the correct draft aft?

$$d = \frac{MCT\ 1\ cm \times L}{TPC \times a}$$

$$= \frac{140 \times 150}{20 \times 70}$$

$$= 15 \text{ metres forward of the LCF}$$

WORKED EXAMPLE 45

A vessel drawing 6.75 m forward, 7.75 m aft, MCT 1 cm 140 tonnes-metres, TPC 15 has cargo space available in Nos. 2 and 4 holds, 50 m forward and 40 m abaft the centre of flotation which is at amidships. How much cargo should be loaded in each hold if the ship is to complete loading with a mean draft of 8.0 m and trimmed 15 cm by the stern?

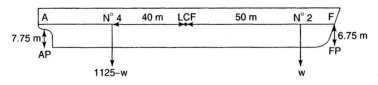

Figure 5.9

Present draft	6.75 m F		7.25 m mean	
	7.75 m A		8.00 m required	
Trim now	1.00 m by the stern ↻		0.75 m sinkage =	75 cm
Required trim	0.15 m by the stern ↻		TPC	×15
Change of trim	0.85 m by the head ↻		To load	1125 tonnes

Let w tonnes be loaded in No. 2 hold

Then (1125 − w) tonnes will be loaded in No. 4 hold

Now the Resultant Moment = The moment to be caused

$$\text{Change of trim} = \frac{\sum (w \times d)}{\text{MCT 1 cm}}$$

So $85 \times 140 = \sum (w \times d)$

$\therefore \sum (w \times d) = 85 \times 140$

Forward Aft   Forward

$50w - 40(1125 - w) = 85 \times 140$

$50w - 45\,000 + 40w = 11\,900$

$90w = 56\,900$

$w = 632.22$ tonnes in No. 2 hold.

and $= 492.78$ tonnes in No. 4 hold.

WORKED EXAMPLE 46

A vessel MCT 1 cm 150 tonnes-metres TPC 20 is drawing 8.4 m F 9.0 m A. She is to discharge 765 tonnes of cargo of which 425 tonnes is discharged from No. 3 hold, the C.G. of which is 6.5 m abaft the centre of flotation. How much should be discharged from Nos 1 and 5 holds, 50 m forward and 40 m abaft the centre of flotation respectively to complete discharging on an even keel?

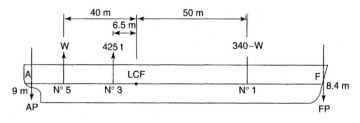

Figure 5.10

Longitudinal stability, i.e. Trim

Present draft	8.40 m F	Total to discharge	= 765 tonnes
	9.00 m A	Already discharged	= 425 tonnes
Trim now	0.60 m by the stern	To discharge from Nos 1 and 5 holds	= 340 tonnes
Required trim	0		
Change of trim	0.60 m by the head ↺		

Let w tonnes be discharged from No. 5 hold.

Then (340 − w) tonnes will be discharged from No. 1 hold.

Now: The resultant moment = The moment to be caused.

      Forward   *Forward*   *Aft*   *Forward*

      40 w   + (425 × 6.5) −   50(340 − w) = 60 × 150
      40 w   +   2762.5   − 17 000 + 50 w = 9000
                                    90 w = 23 237.5
                                      w = 258.2 tonnes from No. 5 hold
                                      and 81.8 tonnes from No. 1 hold

    It will be noted that in the two previous examples w tonnes are loaded at or discharged from the end where the Greater Moment is to be caused. Students will probably find it convenient to do likewise, although so long as the smaller moment is taken from the greater moment it really does not matter at which end w is placed. A *sketch* will always assist in finding a solution to a trim problem.

WORKED EXAMPLE 47

A vessel of 5080 tonnes light displacement is at present floating at a mean draft of 7 m and is trimmed 60 cm by the stern, her deadweight being 5720 tonnes. Her length is 150 m, GM_L 200 m, TPC 25. In order to pass over a bar her after draft is not to exceed 7.2 m. Calculate the minimum amount of water to put in the forepeak tank 60 m forward of the centre of flotation (which is amidships) to achieve the required draft.

$$W = Lwt + Dwt = 5080 + 5720 = 10\,800 \text{ tonnes.}$$

$$\text{MCT 1 cm} = \frac{W \times GM_L}{100L} = \frac{10\,800 \times 200}{100 \times 150} = 144 \text{ tonnes-metres}$$

$$\text{mean bodily sinkage} = W/TPC.$$

Let w tonnes be loaded in the forepeak.

This will cause a mean bodily sinkage of $\dfrac{w}{25}$ cm

and change of trim of $\dfrac{60w}{144}$ cm

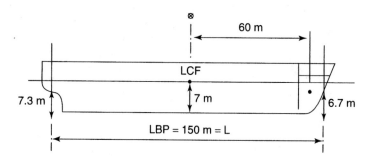

Figure 5.11

giving a trim ratio aft of $\dfrac{60w}{144 \times 2}$ (as LCF is at amidships) ---------- (I)

$$
\begin{array}{rl}
\text{Mean draft} = & 7.00\,\text{m} \\
\text{1/2 present trim} = & +0.30\,\text{m} \\ \hline
\text{Present draft aft} = & 7.30\,\text{m} \\
\text{Sinkage} = & \dfrac{w}{2500}\,\text{m} \\ \hline
 & 7.3 + \dfrac{w}{2500}\,\text{m} \\
\text{Required draft} = & 7.2\,\text{m} \\ \hline
\text{Change required} = 0.1 + & \dfrac{w}{2500}\,\text{m} = \left(10 + \dfrac{w}{25}\right)\,\text{cms} \quad \text{------- (II)}
\end{array}
$$

Now equation (II) = equation (I)
Change required = trim ratio aft.

$$10 + \dfrac{w}{25}\,\text{cm} = \dfrac{60w}{288}\,\text{cm}$$

$$\dfrac{5w}{24} - \dfrac{w}{25} = 10$$

$$w = 59.4\,\text{tonnes}$$

Another method by which the above type of problem can be solved is shown in the next example.

WORKED EXAMPLE 48

A vessel length 120 m, MCT 1 cm 120 tonnes-metres, TPC 15 is drawing 6.8 m F, 7.1 m A. It is required to bring the after draft to 6.8 m by pumping out water from the after peak tank whose centre of gravity is 50 m abaft the LCF which is 4 m abaft amidships What is the minimum quantity of water to be discharged?

Longitudinal stability, i.e. Trim

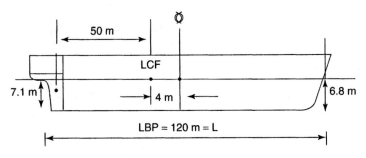

Figure 5.12

Let y cm be the bodily rise when the water is discharged.

The amount of water pumped out will be 15y tonnes.

$$
\begin{array}{rr}
\text{Draft aft} = 7.10 \text{ m or} & 710 \text{ cm} \\
\text{Bodily rise} = & y \text{ cm} \\
\hline
& (710 - y) \text{ cm} \\
\text{Required draft} & 680 \text{ cm} \\
\hline
\text{Change required} & (30 - y) \text{ cm}
\end{array}
$$

Change of trim to be caused is $(30 - y)\dfrac{120}{56}$

This requires a moment of $(30 - y)\dfrac{120}{56} \times 120$ tonnes-metres.

The moment caused by discharging from A.P. is $15y \times 50$ tonnes-metres.

$$\text{Then } 15y \times 50 = (30 - y)\dfrac{120}{56} \times 120$$

$$\text{Whence } y = 7.659$$

$$\text{Water to pump out} = 15y$$

$$= 114.9 \text{ tonnes.}$$

TRUE MEAN DRAFT

The draft at the centre of flotation (LCF) is the *true* mean draft. This does not change if the vessel is trimmed by shifting weights fore and aft. The mean draft amidships changes with trim unless the centre of flotation is at amidships. There is a correction to apply to the apparent mean draft in order to obtain the true mean draft and this is illustrated in the next example. The correction is sometimes known as the correction for layer.

With full form ships such as supertankers and fine form vessels such as Container Ships, due to LCF being up to 3% L for'd and 3% L aft of amidships respectively, this true mean draft concept becomes very important to consider. The correction for draft can mean large differences in the displacement.

The *true mean draft* is measured immediately below LCF position on the waterline. The *average draft* is the draft measured at amidships, i.e. $\dfrac{\text{draft}_{AP} + \text{draft}_{FP}}{2}$.

WORKED EXAMPLE 49

A vessel whose length is 150 m, TPC of 20 is drawing 6.8 m F and 8.0 m A. Her LCF is 5 m aft of amidships. How much cargo can be loaded if she is to complete loading with an even keel draft of 7.70 m?

Present drafts: $\left.\begin{array}{l} F = 6.8\,m \\ A = 8.0\,m \end{array}\right\}$ average of 7.4 m @ amidships

Trim = 1.2 m by the stern

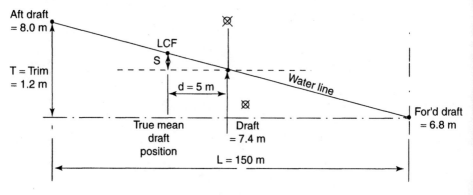

Figure 5.13

Let S be the difference in cm between the draft under LCF position and the draft at amidships. The former draft is the true mean draft.

Let d = distance of LCF from amidships

Let T = trim in cm.   Let L = ship's length in metres.

By similar triangles, $\dfrac{S}{d} = \dfrac{T}{L}$. So $S = \dfrac{T \times d}{L} = \dfrac{120 \times 5}{150}$

Hence S = 4 cm or 0.04 m.

Longitudinal stability, i.e. Trim

$$\text{average draft, @ amidships} = 7.40 \text{ m}$$
$$+ \text{ Correction of S} = \underline{0.04} \text{ m}$$
$$\text{true mean draft} = 7.44 \text{ m @ LCF position}$$
$$\text{required draft} = 7.70 \text{ m @ LCF position}$$
$$\text{Required bodily sinkage} = \underline{0.26} \text{ m or 26 cm, say } \delta d$$
$$\text{Cargo to load} = \delta d \times \text{TPC} = 26 \times 20$$
$$= 520 \text{ tonnes.}$$

BILGING AN END COMPARTMENT

The effects of bilging a midship compartment were considered in Worked Examples 8 and 9. If an end compartment is bilged there will still be a loss of buoyancy and therefore a sinkage. The loss of buoyancy at the end of a vessel causes a shift of the centre of buoyancy away from the end and this shift is the trimming lever on which the total buoyancy acts to cause a trimming moment and therefore a change of trim.

Loss of waterplane area at the end also causes a movement of the centre of flotation so that the change of draft at each perpendicular due to the change of trim is not equal. The following example illustrates this.

WORKED EXAMPLE 50

A box shaped vessel 170 m long and 15 m beam, is floating on an even keel in salt water at a draft of 4 metres. A forward end compartment, 10 m long and the full breadth of the ship is bilged. Calculate the new drafts if the MCT 1 cm is 100 tonnes-metres.

$$\text{Bodily sinkage} = \frac{\text{The volume of lost buoyancy}}{\text{The area of the intact waterplane}} \text{ (see page 13)}$$

$$= \frac{15 \times 10 \times 4}{(170 - 10) \times 15} = 0.25 \text{ m}$$

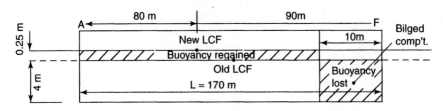

Figure 5.14

The buoyancy lost at the end is transferred to the intact part of the waterplane. We can for practical purposes treat the buoyancy lost forward as a weight loaded. It should be noted that the position of the LCF has changed. This is a very important change.

Buoyancy lost = $15 \times 10 \times 4 \times 1.025 = 615$ tonnes

The moment caused = $615 \times 85 = 52\,275$ tonnes-metres

Change of trim = $\dfrac{\text{Moment caused}}{\text{MCT 1 cm}} = \dfrac{52\,275}{100}$

= 522.75 cm by the head ↻

Trim ratio for'd = $\dfrac{522.75 \times 90}{170} = 276.75$ cm increase

Trim ratio aft = $\dfrac{522.75 \times 80}{170} = 246.00$ cm decrease

	A	F
Old draft	4.000 m	4.000 m
Sinkage	0.250 m	0.250 m
	4.250 m	4.250 m
Trim ratio	−2.460 m	+2.768 m
New end drafts	1.790 m	7.018 m
	say 1.79 m	say 7.02 m

CHANGE OF DRAFT WHEN PASSING BETWEEN WATER OF DIFFERENT DENSITIES

When a vessel passes from water of greater density to water of lesser density there will be a bodily sinkage plus a *slight* change of trim. Worked Example 51 shows the principles involved.

In addition to the sinkage there will be a shift in the position of the centre of buoyancy. The centre of buoyancy will move towards the geometrical centre of the extra layer of buoyancy. If the centre of buoyancy is not vertically below the centre of the layer there will be a fore and aft movement of the centre of buoyancy as well as a vertical movement. The centre of flotation can, for practical purposes, be considered as the centre of the layer. See Figure 5.15.

The change in the fore and aft position results in a lever between the forces of gravity and buoyancy which in turn causes a trimming moment, as the vessel is only in longitudinal equilibrium when the centres of gravity and buoyancy are on the same vertical line.

$$BB_1 = \dfrac{v}{V} \times \text{distance } BC_f$$

where v is the volume of layer of additional buoyancy

V is volume displaced by vessel

The horizontal part of the shift BB_1 is the length of the trimming lever and can be found by multiplying the volume of the layer by the fore and aft distance between the centres of flotation and buoyancy and dividing by the volume of displacement. The trim is always in the *opposite* direction to the shift of the centre of buoyancy.

Longitudinal stability, i.e. Trim

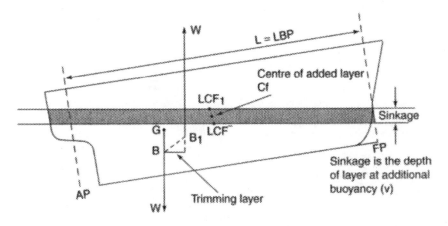

Figure 5.15

Passing from water of less density to that of greater density also affects draft, but in the opposite sense to that described above.

WORKED EXAMPLE 51

A box shaped vessel 200 metres length on the waterline has a breadth of 25 metres and is floating in salt water at drafts of 9.0 metres forward and 11.0 metres aft. Calculate her drafts on passing into fresh water.

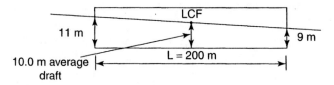

Figure 5.16

$$\text{Original draft} \quad \left.\begin{array}{l} 9.0\,\text{m F} \\ 11.0\,\text{m A} \end{array}\right\} 10.0\,\text{m average}$$

$$\text{Trim} \quad \underline{2.0\,\text{m by}} \quad \text{stern} = 200\,\text{cm by stern} \circlearrowleft$$

$$\begin{aligned}
\text{Volume of displacement in salt water} &= 200 \times 25 \times 10 = 50\,000\,\text{m}^3 \\
\text{Volume of displacement in fresh water} &= 50\,000 \times \frac{1.025}{1.000} = 51\,250\,\text{m}^3 \\
\text{Volume of layer} &= \underline{12\,50\,\text{m}^3}
\end{aligned}$$

$$\text{Mean bodily sinkage} = \frac{\text{Volume of layer}}{\text{Area of layer}} = \frac{1250}{200 \times 25} = 0.25\,\text{m}$$

(assuming $GM_L = BM_L$)

$$\text{MCT 1 cm} = \frac{W \times GM_L}{100\,L} = \frac{50\,000 \times 1.025 \times 200 \times 200}{100 \times 200 \times 12 \times 10} = 854\,\text{t.m}$$

Original distance of B abaft amidships $= \dfrac{\text{Old trimming moment}}{\text{Displacement}} = \dfrac{854 \times 200}{51\,250}$ m

(CF at midlength as waterplane rectangular) $= 3.333$ metres.

Shift of B horizontally $= \dfrac{1250}{51\,250} \times 3.333 = 0.0813$ metres.

Change of trim $= \dfrac{\text{Displacement} \times \text{Lever}}{\text{MCT 1 cm}} = \dfrac{51\,250 \times 0.0813}{854}$

$= 4.88$ cm by stern. ↻

i.e. a very *slight* change of trim

	Forward	Aft
Original drafts in S.W	11.000 m	9.000 m
Mean bodily sinkage	0.250 m	0.250 m
	11.250 m	9.250 m
Trim ratio	+ 0.024 m	−0.024 m
Drafts in fresh water	11.274 m	9.226 m
	say 11.27 m	say 9.23 m

In practice, the slight change of trim is ignored. However for exam purposes it must be understood and calculated in density problems.

CHAPTER SIX

Drydocking

When a vessel is to drydock it is usual to have her trimmed by the stern. This enables the stern to be set on the blocks, and then used as a pivot to align the keel along the blocks. If the vessel is not trimmed by the stern greater skill will be needed in the manipulation of the mooring tackles forward and aft.

When the vessel first touches the blocks, the whole of the vessel's weight is being supported by the buoyancy of the water. As water is pumped from the dock, part of the buoyancy of the water is transferred to the keel blocks, this is called the UPTHRUST (P). It will be shown that this upthrust causes a loss of GM. Any loss in GM will of course be a loss in the ship's stability.

To prevent the loss of GM causing instability:

(a) The vessel should have an adequate GM.

(b) Free surface in the tanks should be at a minimum.

(c) The vessel should not be trimmed too much by the stern. In practice, the trim prior to drydocking will be 0.30 m to 1.00 m by the stern.

The critical time during the drydocking operation is just before the vessel takes the blocks fore and aft. Until she is on the blocks throughout her length, the shores cannot be set up tight, and it should be ensured that she still has positive stability at the time. When the vessel has taken the blocks overall pumping of the dock ceases until the shores have been set up tight.

When placing the shores on the ship's side, it should be seen that the end of the shore is over a frame and, if possible, at the intersection of a frame and a deck stringer, as the greater strength is at this point.

A critical time also occurs when refloating. The distribution of weight should be similar to that when the vessel drydocked. Tank soundings should be taken as soon as the vessel has taken the blocks overall. Similar soundings should be obtained before refloating.

If any tank plugs in the bottom of the vessel have been removed the mate should see that they have been replaced before the dock is flooded.

Whilst in drydock a fire hose from the shore should be connected to the ship's fire main. All discharges overside should be stopped.

TO FIND THE UPTHRUST 'P'

The Ram bow in Figure 6.1 has similar advantages as the Bulbous bow; and is cheaper to construct into the for'd end of a ship.

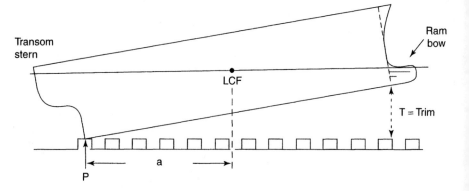

Figure 6.1

In the diagram above

    P is the upthrust

    a is the distance of the sternpost from the centre of flotation (LCF)

    T is the trim

In order to have the vessel on the blocks fore and aft, a moment has to be caused to change the trim T centimetres. This moment could be caused by loading a weight forward of the LCF, discharging a weight from abaft the LCF, shifting a weight from aft to forward or pushing up at the after part of the ship.

The last named method is virtually what the upthrust does. The moment it causes is $P \times a$.

Then: The moment caused = The moment to be caused.

$$\text{i.e. } P \times a = T \times \text{MCT 1 cm}$$

$$\text{so } P = \frac{T \times \text{MCT 1 cm}}{a}$$

Note: This is the upthrust at the instant the stern takes the blocks.

The upthrust at any time prior to this can be found by substituting the change of trim up to that time for T in the equation above.

When a vessel is on the blocks and on an even keel, the upthrust will increase by an amount equal to the TPC for every centimetre of water that is pumped out of the dock.

$$\text{i.e. } P = \text{TPC} \times \text{the number of centimetres the water has fallen.}$$

TO FIND THE LOSS OF GM

Consider the vessel heeled to an angle $\theta°$

$\theta°$ can be the smallest angle that there is (e.g. 1 second of arc)

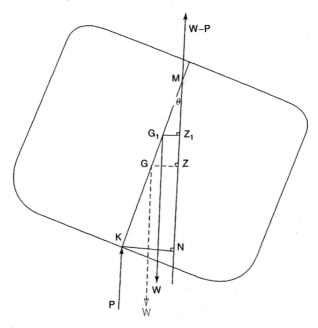

Figure 6.2

Taking moments about M (considering it to be fixed)

$$W \times GZ - P \times KN = W \times G_1Z_1$$

or $W \times GM \sin \theta - P \times KM \sin \theta = W \times G_1M \sin \theta$

then $W(GM - G_1M) = P \times KM$

i.e. $GM - G_1M = \dfrac{P \times KM}{W}$

(or the loss of GM)

Alternatively if P is considered as a weight discharged from the keel, the shift of G or loss of GM can be found by taking moments about G.

Then The loss of $GM = \dfrac{P \times KG}{W - P}$

Either of the above approximate formulae may be used in working out problems.

Whilst there is a difference in the effective GM and therefore the GZ there will be practically *no difference in the Righting Moment* which is the important quantity in ship stability. The reason that the righting moments will be the same is because the different GZs will be multiplied by different displacements – using the first formula the displacement will be W, whereas it will be W − P when using the second formula. This is shown in Worked Example 53.

WORKED EXAMPLE 52

A vessel of 3000 tonnes displacement, KM (considered fixed) 4.5 metres MCT 1 cm 80 tonnes-metres is trimmed 50 cm by the stern. The sternpost is 40 metres abaft the centre of flotation. Calculate the loss of GM at the instant the stern takes the blocks.

$$P = \frac{C \text{ of } T \times MCT \ 1 \text{ cm}}{a} = \frac{50 \times 80}{40} = 100 \text{ tonnes.}$$

$$\text{Loss of GM} = \frac{P \times KM}{W} = \frac{100 \times 4.5}{3000} = 0.15 \text{ metres.}$$

WORKED EXAMPLE 53

A vessel of 4000 tonnes displacement TPC 12, KM (considered fixed) 4.2 m, KG 3.2 m is on an even keel. She enters drydock and is set down on to the blocks. What is her effective GM when a further 0.45 metres of water is pumped from the dock? What is the value of the final righting moment?

$$P = TPC \times \text{cm of water pumped out}$$
$$= 12 \times 45 = 540 \text{ tonnes.}$$

$$\text{Loss of GM} = \frac{P \times KM}{W} = \frac{540 \times 4.2}{4000} = 0.567 \text{ m}$$

$$\text{Old GM} = \underline{1.000 \text{ m}}$$
$$\text{Effective GM} = \underline{0.433 \text{ m.}}$$

Alternatively:

$$\text{Loss of GM} = \frac{P \times KG}{W - P} = \frac{540 \times 3.2}{3460} = 0.499 \text{ m}$$

$$\text{Old GM} = \underline{1.000 \text{ m}}$$
$$\text{Effective GM} = \underline{0.501 \text{ m}}$$

Righting moment $= 0.433 \times 4000 \qquad = 1732 \text{ t.m}$ } This verifies the statement made
$\qquad$ or $\quad = 0.501 \times (4000 - 540) = 1733 \text{ t.m}$ } prior to Worked Example 52.

CHAPTER SEVEN

Water and Oil Pressure

Pressure is force per unit area, and in this section the force is due to a head of liquid.

Consider an area of one square metre 1 metre under the surface of fresh water. The volume of water over the area is $1\,m^3$. From the previous work in Chapter one it is known that this weighs 1 tonne or 1000 kgf and therefore the pressure is $1000\,kgf/m^2$. If the depth of fresh water over the same area is increased to 2 metres the volume of water will be increased to $2\,m^3$ which weighs 2 tonnes or 2000 kgf so the pressure now is $2000\,kgf/m^2$. If the depth of water is further increased the pressure will also increase, therefore pressure can be said to vary directly with depth. The depth in the case of horizontal areas is also the pressure head, further mention of pressure head is made on the next page.

If the water had been other than fresh its weight per cubic metre would have been greater than 1 tonne and if a liquid with a density less than that of fresh water had been used its weight per cubic metre would have been less than 1 tonne. The pressure would also have been different from that for fresh water so pressure varies directly with the density, or in practical terms the relative density.

Increasing the area under water will increase the volume of water over the area. Suppose the depth of water remains constant at 1 metre whilst the area under the water is increased to

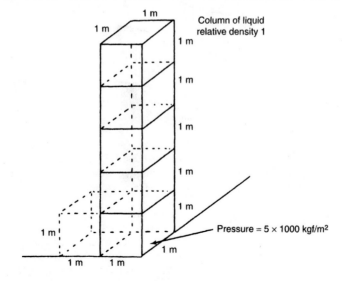

Figure 7.1

5 metres. The weight of water over the area is 5 tonnes or 5000 kgf, but the weight per unit area is only 5000/5 kgf/m² or 1000 kgf/m². So the pressure is quite independent of the area under pressure, but the weight over the whole area depends on the area under pressure. The total weight of liquid over an area is called the thrust.

To summarise.

$$\text{The thrust on an area} = A \times h \times \rho \text{ tonnes} = P$$

Where A is the area under pressure in square metres
h is the pressure head in metres
ρ is the density of the liquid in tonnes per cubic metre

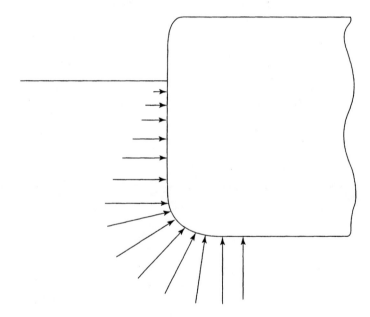

Figure 7.2

The thrust always acts at right angles to the immersed surface and for any set depth the thrust upwards, downwards, sideways and any other direction will be exactly the same. It can be shown that the pressure head to be used for obtaining the thrust on any area is the depth of the geometrical centre of the area below the surface of the liquid.

Centre of pressure of an area is the point on the area where the total thrust could be considered to act. It will not be necessary to calculate this point. However, it should be borne in mind that if a bulkhead requires shoring due to flooding on the other side, the most effective point to place the shore is at the centre of pressure. For rectangular areas vertically placed (ordinary bulkheads) the centre of pressure is $\frac{2}{3}$ depth below the surface. For triangular areas (collision bulkheads) the centre of pressure is at half depth.

Water and oil pressure

WORKED EXAMPLE 54

A double bottom tank 20 metres long, 15 metres wide is being tested with a head of 2.5 metres of salt water. Calculate the thrust on the tanktop plating.

$$\text{Thrust} = A \times h \times \rho_{SW} = P$$
$$P = (20 \times 15) \times 2.5 \times 1.025$$
$$P = 768.75 \text{ tonnes.}$$

WORKED EXAMPLE 55

A rectangular lock gate 20 metres wide, 15 metres deep, has 6 metres of water R.D. 1.025 on one side, whilst on the other side there is 10 metres of water R.D. 1.012. Calculate the thrust on each side, the resultant thrust and the Centre of Pressure above base at which this resultant thrust acts.

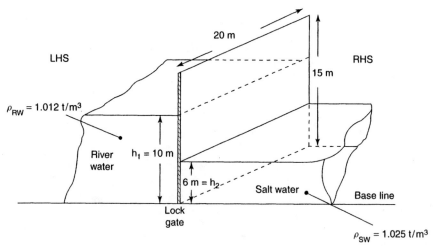

Figure 7.3

Salt water side:

$$P = \text{Thrust} = A \times h_2 \times \rho_{SW}$$
$$P = 20 \times 6 \times 3 \times 1.025$$
$$P = 369 \text{ tonnes.}$$

Other side:

$$P = \text{Thrust} = A \times h_1 \times \rho_{RW}$$
$$P = 20 \times 10 \times 5 \times 1.012$$
$$P = 1012 \text{ tonnes.}$$

	Thrust	Centre of Pressure (base)	Moment of Pressure
LHS: River water	1012	10/3 = 3.33 m	3370
RHS: Salt water	−369	6/3 = 2.00 m	−738
	643 t		2632 t·m

Resultant thrust = 643 tonnes

$$\text{Centre of Pressure} = \frac{\sum \text{moments}}{\sum \text{thrusts}} = \frac{2632}{643} = 4.09 \text{ m above base from LHS}$$

Note:

For the river water, centre of pressure acts at $2/3\, h_2$ from top of liquid. This is the same as $1/3\, h_2$ above the base i.e. 10/3 m, or 3.33 m.

For the salt water, centre of pressure acts at $2/3\, h_1$ from top of liquid. This is $1/3\, h_1$ above base i.e. 6/3 m, or 2 m

Final solution can be portrayed as follows.

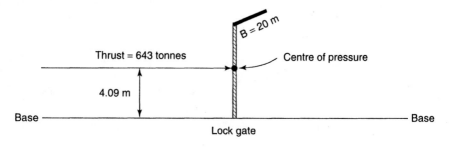

Figure 7.4

WORKED EXAMPLE 56

Calculate the upthrust 'P' on the bottom plating of a box-shaped vessel 75 m long, 10 m wide, 7 m deep, when floating at a draft of 5 m in fresh water.

$$\text{Thrust } P = A \times h \times \rho$$
$$P = (75 \times 10) \times 5 \times 1.000$$
$$P = 3750 \text{ tonnes}$$

Water and oil pressure   115

It should be noted that this thrust on the bottom plating is the same as the vessel's displacement. This is to be expected because the principle of flotation is that the upthrust due to buoyancy is equal to the vessel's displacement.

Now consider an example where the shape of the bulkhead changes to being *triangular* and liquid pressure is *water* on one side together with *oil* on the other side.

WORKED EXAMPLE 57

A main transverse bulkhead is triangular shaped. It has liquid pressure on both sides as shown in Fig. 7.5. Calculate the resultant thrust and its centre of pressure above base.

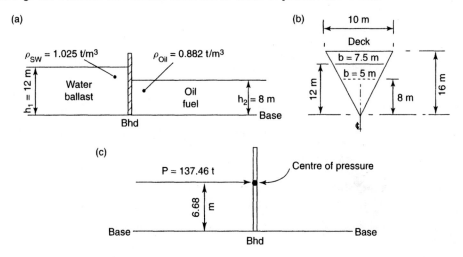

Figure 7.5

Consider LHS only

$$P = \text{Thrust} = A \times h_1 \times \rho_{SW}$$

$$= \frac{12 \times 7.5}{2} \times \frac{12}{3} \times 1.025$$

$$= 184.50 \text{ tonnes, acting @ } h_{1/2} \text{ from top of water ballast is 6 m above base.}$$

Consider RHS only

$$P = \text{Thrust} = A \times h_2 \times \rho_{OIL}$$

$$= \frac{8 \times 5}{2} \times 8/3 \times 0.882$$

$$= 47.04 \text{ tonnes, acting @ } h_{2/2} \text{ from top of oil is 4 m above base.}$$

Location	Thrust	Centre of Pressure (base)	Moment of Pressure
LHS	184.50	6 m	1107.00
RHS	−47.04	4 m	−188.16
	$137.46 = \sum_1$		$918.84 = \sum_2$

Resultant Thrust = 137.46 tonnes from LHS.

Centre of Pressure = $\dfrac{\sum_2}{\sum_1} = \dfrac{918.84}{137.46} = 6.68$ m above base

Final solution can be portrayed as shown in Fig. 7.5(c).

WORKED EXAMPLE 58

A cylindrical pressure vessel for the carriage of compressed gases has a length of 12 m and a diameter of 7 m. It is fitted vertically in a ship. The maximum pressure to which the top is to be subjected is 0.6 kgf/cm^2.

If the top of the cylinder is to be water tested with salt water to this pressure how far up the stand pipe would the water be? What would be the thrust on the cylinder top at this time?

If the water is fresh 1 cm^3 weighs 1 gramme and the pressure or unit area is 1 gf/cm^2 when the column is 1 cm high. If a pressure of 0.6 kgf (600 gf) is required on unit area (cm^2) a column 600 cm high is required.

Because the water is salt water, the height will be $\dfrac{600}{1.025} = h$

Therefore h = 585.4 cms or 5.854 m.

Thrust P on tank top = pressure head × area × density

$$\therefore \; P = h \times \pi r^2 \times \rho = \dfrac{5.854 \times 22 \times 3.5 \times 3.5 \times 1.025}{7} = 231 \text{ tonnes}$$

WORKED EXAMPLE 59

A collision bulkhead bounding the Fore Peak tank is 24 m in depth. Starting from the top, it has equally spaced ordinates: 34.0, 33.3, 32.5, 30.8, 26.5, 17.3 and 6.2 m. Calculate the thrust and its Centre of Pressure of above base when the tank is full of water RD of 1.016.

Water and oil pressure

No. of ord.	ord.	SM	Area ftn	lever	Moment ftn	lever	Inertia ftn
1	34.0	1	34.0	0	0	0	0
2	33.3	4	133.2	1	133.2	1	133.2
3	32.5	2	65.0	2	130.0	2	260.0
4	30.8	4	123.2	3	369.6	3	1108.8
5	26.5	2	53.0	4	212.0	4	848.0
6	17.3	4	69.2	5	346.0	5	1730.0
7	6.2	1	6.2	6	37.2	6	223.2
			$483.8 = \sum_1$		$1228.0 = \sum_2$		$4303.2 = \sum_3$

$$\text{Area of Bulkhead} = \frac{1}{3} \times \sum_1 \times h$$

$$= \tfrac{1}{3} \times 483.8 \times 4$$

$$= 645.07 \, m^2.$$

$$\bar{y} = \text{CG } below \text{ the surface} = \frac{\sum_2}{\sum_1} \times h$$

$$= \frac{1228}{483.8} \times 4$$

$$= 10.153 \, m.$$

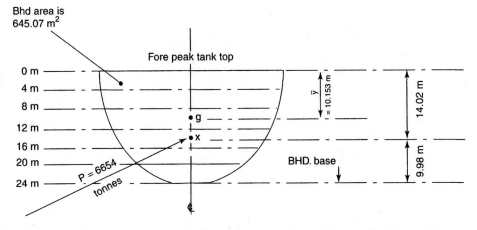

Figure 7.6 This figure shows clearly that the Centre of Gravity is different to the Centre of Pressure. Confusion over these two values often exists.

$$\text{Thrust 'P' on Bulkhead} = A \times \bar{y} \times \rho$$
$$= 645.07 \times 10.153 \times 1.016$$
$$= 6654 \text{ tonnes.}$$

$$\text{Centre of Pressure for 'P'} = \frac{\Sigma_3}{\Sigma_2} \times h = x.$$

$$\text{C. of Pr.} = \frac{4303.2}{1228} \times 4 = x.$$

C. of Pr. = 14.02 m below surface = x.

C. of Pr. = 9.98 m above base.

Note:

A *sketch* will always be useful when working with problems of this type.

Procedure:

1. Make a sketch from the given information in question.
2. Use a table and Simpson's rule to obtain Σ_1, Σ_2 and Σ_3.
3. Calculate area 'A' of bulkhead.
4. Calculate VCG or $\bar{y}$ below top of liquid.
5. Calculate Thrust 'P' using $P = A \times \bar{y} \times \rho$ formula.
6. Finally calculate centre of Pressure *below* surface of the liquid and modify it to a distance vertically *above* base.

CHAPTER EIGHT

Free Surface Effects

Whenever there is a surface of liquid which is free to move, there is a loss of effective GM. This loss takes place irrespective of the position of the free surface in the ship. Any loss in GM decreases the ship's stability

As can be seen from the diagram below, the movement of liquid in the tank caused a shift, to the low side, of the ship's centre of gravity. The tank is shown to be half full of water and for the angle of heel illustrated, its centre of gravity will shift from g to g_1.

If the tank is more than half full, then for the same angle of heel, the movement of its centre of gravity will be less. Conversely if the tank is less than half full there will be a greater movement of its centre of gravity. Any tank partially filled is called a 'slack tank'.

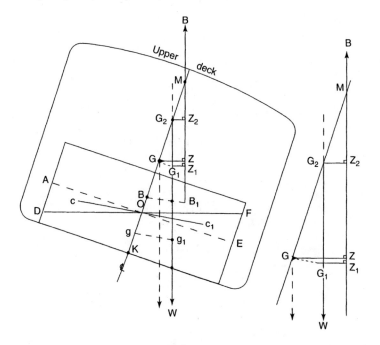

Figure 8.1

When G shifts to G_1 there is a reduction in the righting lever from that which there would be if there were no free surface (B_1 remaining fixed as the angle of heel and displacement do not change). The effective righting lever now is G_1Z_1. The righting lever is the perpendicular distance between the forces of gravity and buoyancy and can be drawn at any convenient point. When considering free surface effect it is convenient to draw the righting lever from the point where the line of action of gravity cuts the centre line of the vessel; this point is G_2. G_2Z_2 is the equal to G_1Z_1 so G_2M is the effective GM. G_2 is the virtual position of the ship's centre of gravity (note the actual position is at G_1) and GG_2 is the virtual loss of GM due to the free surface effect.

Let l be the length of the free surface and b be the breadth of the free surface then $\frac{b}{2} =$ 1/2 breadth $= AO = OE = y$.

c and c_1 are centres of gravity of wedges of liquid in original position (AOD) and transferred position (EOF) when the vessel is heeled $\theta°$ (a small angle).

Let ρ_t be density of liquid in tank.

Let ρ_s be density of water in which vessel is floating.

$$cO = C_1O \simeq \frac{2}{3}y \qquad \text{so } cc_1 \simeq \frac{4}{3}y$$

$$\text{Volume of wedge AOD} = y\tan\theta° \times \frac{y}{2} \times 1$$

$$\text{Weight of wedge} = \text{volume} \times \rho_t$$

Moment caused by transferring wedge from c to $c_1 = 1/2y^2 \tan\theta° \times 1 \times \rho_t \times \frac{4}{3}y$

When expressed in circular measure $\tan\theta° = \theta_R$

$$\text{So moment} = \frac{2}{3}y^3 \times 1 \times \theta_R \times \rho_t$$

But this moment causes a shift of G to G_1

$$\text{such that } GG_1 = \frac{\frac{2}{3}y^3 \times 1 \times \theta_R \times \rho_t}{V \times \rho_s}$$

Where $V \times \rho_s$ = Vessel's displacement

$$\text{Also } GG_1 = GG_2 \times \theta_R$$

$$\therefore GG_2 \times \theta_R = \frac{\frac{2}{3}y^3 \times 1 \times \theta_R \times \rho_t}{V \times \rho_s}$$

as $\frac{2}{3}y^3 \times 1$ = moment of inertia free of surface (i)

$$GG_2 = \frac{i}{V} \times \frac{\rho_t}{\rho_s} \quad \text{or} \quad \frac{i}{W} \times \rho_t$$

Free surface effects

where y = 1/2 breadth of rectangular free surface $i = \frac{2}{3} \times \left(\frac{b}{2}\right)^3 \times 1 = \frac{lb^3}{12}$

If the free surface is divided *longitudinally* it can be shown that the virtual loss of GM (GG_2) varies inversely as the square of the number (n) of compartments.

$$\text{Hence } GG_2 = \frac{i}{W} \times \frac{\rho_t}{n^2} \text{ metres.}$$

To remove these free surface effects (FSE) simply fill the tank up to full capacity or completely drain the tank.

WORKED EXAMPLE 60

A vessel displacing 8000 tonnes in salt water has a double bottom tank 20 metres long and 16 metres wide, partly full of sea water. If the ship's KM is 7 metres and the KG is 6 metres, calculate the effective GM if (a) the D.B. is undivided, (b) there is a centre line division, (c) there is a centre line division and two watertight side girders.

(a) $\quad \text{Loss of GM} = \frac{i}{W} \times \rho_t = \frac{lb_1^3 \times 1.025}{12 \times W}$

$\quad\quad\quad\quad\quad\quad\quad = \frac{20 \times 16 \times 16 \times 16 \times 1.025}{12 \times 8000} = 0.875 \text{ m.}$

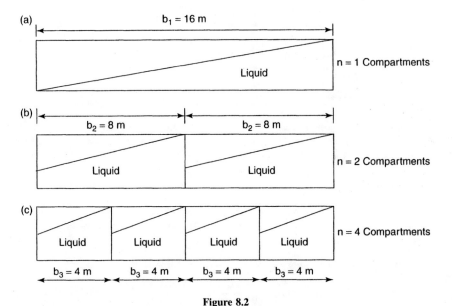

Figure 8.2

Loss in GM = 0.875 m = FSE
GM = 1.000 m
Effective GM = $\overline{0.125\,m}$ Note the use of *three* breadths, b_1, b_2 & b_3. The suffix is very important.

This is *not* acceptable!! It is below D.T$_P$. minium value of 0.15 m!!

(b) Loss of GM = $2 @ \dfrac{i}{W} \times \rho_t = 2 @ \dfrac{lb_2^3 \times 1.025}{12 \times W}$ or $\dfrac{lb_1^3}{12} \times \dfrac{\rho_t}{W} \times \dfrac{1}{n^2}$

$= 2 @ \dfrac{20 \times 8 \times 8 \times 8 \times 1.025}{12 \times 8000}$ or $\dfrac{20 \times 16^3 \times 1.025}{12 \times 8000 \times 2^2} = 0.2187\,m.$

Loss in GM = 0.2187 m i.e. $\dfrac{1}{n^2} \times 0.875\,m$ = FSE
GM = 1.0000 m
Effective GM = $\overline{0.7813\,m}$ say 0.78 m.

(c) Loss of GM = $4 @ \dfrac{i}{W} \times \rho_t$

$= 4 @ \dfrac{lb_3^3 \times 1.025}{12 \times W}$ or $\dfrac{lb_1^3 \times \rho_t}{12 \times W} \times \dfrac{1}{n^2}$

$= 4 @ \dfrac{20 \times 4 \times 4 \times 4 \times 1.025}{12 \times 8000}$ or $\dfrac{20 \times 16^3 \times 1.025}{12 \times 8000 \times 4^2} = 0.0547\,m$

Loss in GM = 0.0547 m i.e. $\dfrac{1}{n^2} \times 0.875\,m$ = FSE
GM = 1.0000 m
Effective GM = $\overline{0.9453\,m}$ say 0.95 m.

The foregoing example emphasises the fact that the greater the number of *longitudinal* subdivisions, the smaller is the loss of GM due to free surface. *Transverse* subdivisions will have *no* effect whatsoever on the original loss in GM of 0.875 m.

WORKED EXAMPLE 61

A ship of displacement 1400 tonnes has 2 metres freeboard with her centre of gravity above the waterline. A rectangular deck area 20 m long and 10 m wide is filled to a depth of 0.5 m when the vessel ships a sea. Find the effect on the vessel's righting lever at 5° heel.

Weight of water shipped = $20 \times 10 \times 0.5 \times 1.025 = 102.5$ tonnes.

Rise of G due to above = $\dfrac{w \times d}{(W+w)} = \dfrac{102.5 \times 2.05}{(1400 + 102.5)} = 0.1397$ metres.

Free surface effects

$$\text{Loss of GM due to free surface} = \frac{i}{W_2} \times \rho_t \times \frac{1}{n^2}$$

$$= \frac{20 \times 10 \times 10 \times 10 \times 1.025}{12 \times 1502.5}$$

$$\therefore \text{Loss in GM} = 1.1369\,\text{m} = \text{FSE}.$$

The total effect on the GM is an effective loss of $0.1397 + 1.1369$ metres i.e. $1.2766\,\text{m}$ $\therefore$ the effect is to decrease the GZ at $5°$ by $1.2766 \times \sin 5°$ i.e. 0.111 metres.

As can be seen this is a significant reduction in initial stability and with such possible reductions in mind the 1968 Load Line Rules require that vessels shall have a certain minimum stability and these requirements are given on page 125.

WORKED EXAMPLE 62

A vessel KM $8.0\,\text{m}$, KG $7.0\,\text{m}$, displacing 9171 tonnes in water relative density 1.019 has a double bottom tank, longitudinally divided, $20\,\text{m}$ long, $15\,\text{m}$ wide and $1.2\,\text{m}$ deep, filled with oil relative density 0.95. Calculate the effective GM if half the oil is used.

Weight discharged $= 20 \times 15 \times 0.6 \times 0.95$

$$= 171.0\,\text{tonnes}.$$

$$W_2 = W_1 - w = 9171 - 171$$

$$= 9000\,\text{t}.$$

KG ship $= 7.00\,\text{m}$ No. of spaces $= 2$,
KG oil $= 0.90\,\text{m}$ so $n = 2$.
$d = 6.10\,\text{m}$

$$GG_1 = \frac{w \times d}{W - w} = \frac{171.0 \times 6.1}{9171 - 171}$$

$$= \frac{1043.1}{9000}$$

$$= 0.116\,\text{metres}. \uparrow$$

$$\text{Loss due to free surface} = \frac{i}{W_2} \times \rho_{OIL} \times \frac{1}{n^2}$$

$$= \frac{lb^3}{12} \times \frac{1}{W_2} \times \rho_{OIL} \times \frac{1}{n^2}$$

$$= \frac{20 \times 15 \times 15 \times 15 \times 0.95}{12 \times 9000 \times 4}$$

Loss in GM $= 0.148\,\text{metres} \uparrow = \text{FSE}.$

KM =	8.000 metres
KG =	−7.000 metres
Original GM =	1.000 metres
GG₁ =	−0.116 metres
	0.884 metres
FSE =	−0.148 metres
Effective GM =	0.736 metres say 0.74 m.

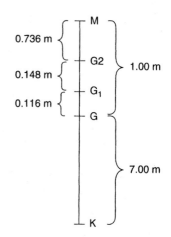

Figure 8.3

Observe in this example how G rose due to FSE and also due to change of loading. It must be emphasised that free surfaces of liquid can be dangerous and they should be kept to a minimum at all times. Remember D.Tp minimum value for GM at all times is 0.15 m, inclusive of free surface effects.

If there is any doubt about a vessel's stability, do not try to rectify a tendency to instability by haphazardly filling double bottom tanks. It must be remembered that before G is brought down due to increased weight in the bottom there will be a virtual rise due to the effect of free surface.

Information about the loss of GM due to free surface must be provided on all vessels whose keels are laid after the 19th November, 1952. This is usually given as a Free Surface Moment, which has to be divided by the displacement of the vessel in order to get the loss of GM.

Note:

Free Surface Moment (FSM) = $i \times \rho_t$ tonnes · m.

where:

    i denotes the moment of inertia of slack tank, in m^4

    ρ_t denotes density of liquid in slack tank, in tonnes/m^3.

CHAPTER NINE

Stability Data

The Load Line Rules require minimum stability conditions as follows:

(a) The area under the curve of Righting Levers (GZ curve) shall not be less than

  (i) **0.055** metre-radians up to an angle of **30** degrees (see Fig. 9.1).

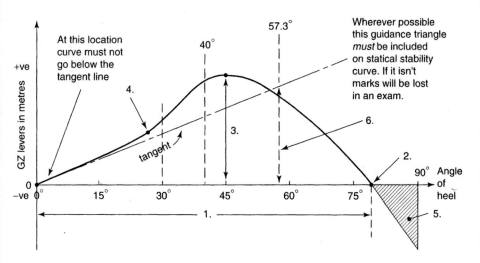

Figure 9.1 Statical Stability Curve

  (ii) **0.09** metre-radians up to an angle of either **40** degrees or the angle at which the lower edges of any openings in the hull, superstructures or deckhouses, being openings which cannot be closed weathertight, are immersed if that angle be less (see Fig. 9.1).

  (iii) **0.03** metre-radians between the angles of heel of **30** degrees and **40** degrees of such lesser angle as is referred to in (ii) (see Fig. 9.1).

(b) The Righting Lever (GZ) shall be at least **0.20** metres at an angle of heel equal to or greater than **30** degrees.

(c) The maximum Righting Lever (GZ) shall occur at an angle of heel not less than **30** degrees.

(d) The initial transverse metacentric height shall not be less than **0.15** metres. In the case of a ship carrying a timber deck cargo which complies with subparagraph (a) by taking into account the volume of timber deck cargo the initial transverse metacentric height shall not be less than **0.05** metres.

The GZ curve referred to above is also known as a curve of Statical Stability and is described as follows

When GZ levers are plotted against angles of heel a curve of Statical Stability is formed. This is illustrated on Fig. 9.1. The information which can be obtained from the curve is detailed below.

1. The range of stability, where all GZ values are positive.

2. The angle of vanishing stability, beyond which the vessel will capsize.

3. The maximum GZ lever, and the angle at which it occurs (30° to 45°).

4. The point at which the deck edge immerses (this is known as the point of contraflexure, i.e. where the shape of the curve changes from concave to convex, when viewed from above).

5. This is negative stability.

6. The approximate GM. A tangent is drawn to the curve at its origin, and a vertical line is drawn from the base line at 57.3°. The distance above the base line of the point where the vertical line and the tangent intersect is the GM. (This is measured on the GZ scale.) See Worked Example 63.

$$GM = \frac{GZ}{\sin\theta°} \text{ or } \frac{GZ}{\theta \text{ Radians}}$$

1 Radian = 57.3°

7. The dynamical stability, this is found by multiplying the area under the curve by the vessel's displacement (see Worked Examples 63 and 64).

Strictly speaking we should be given the GM so as to get the direction of the curve at its origin, but questions may be asked involving the finding of it. See Worked Example 63.

It will be seen that GZ levers are plotted against angles of heel, and it should be understood that when a curve of Statical Stability is drawn, it only applies to the vessel for the *one condition of loading* at which it is drawn. This might be for loaded-arrival, medium-ballast or light-ballast condition.

Effect of position of G on a curve of Statical Stability

It can be seen from Fig. 9.2 that when G is raised to G_1 the GZ lever is reduced by y metres. Also it can be seen that $y = GG_1 \sin\theta°$

Stability data

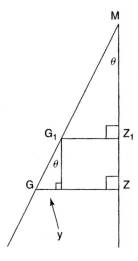

$G_1Z_1 = GZ - y$
$G_1Z_1 = GZ - GG_1 \sin\theta$.

Figure 9.2

Similarly if the centre of gravity was lowered from G_1 to G the GZ lever would be increased by y metres.

$$\text{So the change in } GZ = \pm GG_1 \times \sin\theta°.$$

WORKED EXAMPLE 63

A vessel has the following GZ ordinates taken from GZ Cross Curves of Stability with an *assumed* KG of 8 m @ displacement of 20 550 tonnes.

θ heel	0°	15°	30°	45°	60°	75°	90°
GZ ords(m)	0	1.10	2.22	2.60	2.21	1.25	0.36

When loaded ready for departure her *actual* displacement and KG are 20 550 t and 9.52 m respectively.

(a) Calculate the actual GZ righting levers and draw the Statical Stability curve for this condition of loading

(b) Determine GZ max and the angle at which it occurs

(c) Range of Stability

(d) Dynamical Stability up to 30°

(e) Approximate GM, from Statical Stability curve.

63(a) $GG_1 = 8.00 - 9.52 = -1.52$ m.

Hence $GG_1 \sin\theta = -1.52 \sin\theta$, where $\theta = 0$ to $90°$.

θ.	0°	15°	30°	45°	60°	75°	90°
GZ from Cross Curves (m)	0	1.10	2.22	2.60	2.21	1.25	0.36
$-1.52 \sin\theta$	0	−0.39	−0.76	−1.07	−1.32	−1.47	−1.52
Actual GZ (m)	0	0.71	1.46	1.53	0.89	−0.22	−1.16

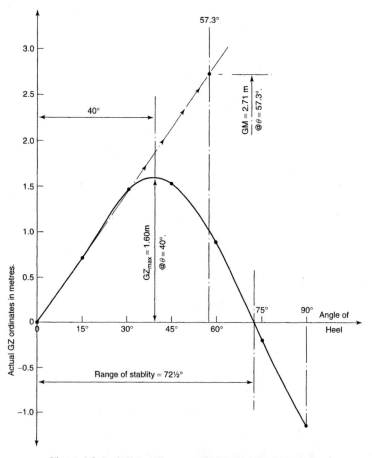

Figure 9.3 Statical Stability curve for Worked Example 63(a)

63(b). $GZ_{max} = 1.60\,m$ @ $\theta = 40°$ from Figure 9.3.
63(c). Range of Stability = $72(1/2)°$. from Figure 9.3.

63(d)

GZ ord	SM	Area ftn
0	1	–
0.71	4	2.84
1.46	1	1.46
		$4.30 = \sum_1$

$$\text{Dynamical Stability} = 1/3 \times CI \times \sum\nolimits_1 \times W$$

$$= 1/3 \times \frac{15°}{57.3°} \times 4.3 \times 20\,550$$

$$= 7711 \text{ metre tonnes.}$$

63(e) Approximately $GM = 2.71\,m$ (extrapolated up to $57.3°$) on Figure 9.3.

Effect of freeboard on a curve of Statical Stability.

The more freeboard that a vessel has, then the greater the angle to which she can be inclined without immersing the deck edge. This is shown in Figure 9.4 with angles θ_1 and θ_2. Breadth is held constant. $\theta_2 > \theta_1$.

Assume for a moment that a ship is being gradually unloaded of cargo after being in loaded-arrival condition. As she has more and more weight discharged then of course has freeboard is increasing from say f_1 to f_2 to f_3.

Figure 9.5 shows the changes that occur in the Statical Stability curves for this ship.

As the freeboard increases the following takes place:

1. GM increases. See plots @ $57.3°$.
2. GZ increases, so Stability increases.
3. Deck edge becomes immersed at higher value of θ, see Fig. 9.5 with θ_1 to θ_3 plots.
4. KB decreases because drafts are less in value.
5. BM increases because volume of displacement is less. This assumes inertia I remains constant.
6. KM increases, as per metacentric Diagram in Fig. 4.3.
7. Range of Stability increases, giving greater dynamical stability.
8. GZ_{max} occurs approximately @ $45°$, then @ $36°$ and finally approximately @ just greater then or equal to $30°$, when at greatest freeboard. This final condition will be when the ship is in Light-ballast or near to Lightship condition. See Fig. 9.5.

To avoid the necessity of carrying a curve for each of the vessel's possible displacements, a set of Cross Curves is drawn. A specimen set of them is illustrated of page 135.

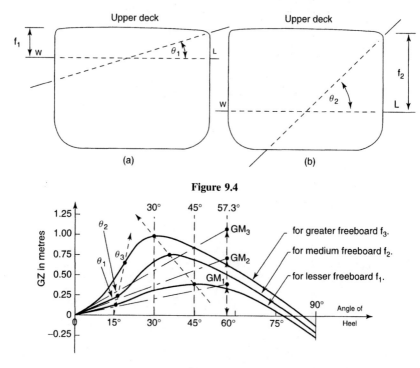

Figure 9.4

Figure 9.5

In order to produce a curve of Statical Stability from the cross curves, the GZ levers for the various angles of heel are taken off for the required displacement. These are then corrected for any difference between the position of the ship's centre of gravity and the centre of gravity for which the curves are drawn. Finally the GZ levers are plotted against the angle of heel. Worked Examples 63 & 64 should make this clear.

Sometimes the curve of Statical Stability is drawn with righting moments instead of GZ levers on the ordinate. The curve can be useful in this form as one can see at a glance how much the vessel will heel when an upsetting moment is applied. The upsetting moment may be caused by transverse movement of weight or, particularly with passenger ships, the effect of a strong wind blowing on the ship's side. Container Ships, carrying four tiers of deck-containers also experience this heeling effect when sailing in high transverse wind conditions.

Effect of beam on a curve of Statical Stability

This must be considered in the design stage. Draft is held constant. $\theta_2 < \theta_1$. See Figures 9.6 and 9.7 on facing page.

It has already been stated that $BM = \dfrac{I}{V}$ and provided that V remains constant BM varies with I. This inertia I is largely dependent on the breadth of the waterplane, so an increase in beam will increase BM. This in turn increases GZ. The effect is most noticeable at the smaller angles of inclination.

An increase in the Beam from b_1 to b_2 will produce the following:

1. GM increases. See plots @ 57.3°.
2. GZ increases, so stability increases.
3. Deck edge becomes immersed at lower value of θ_1, see θ_2 being less than θ_1 in Figs 9.6 and 9.7.

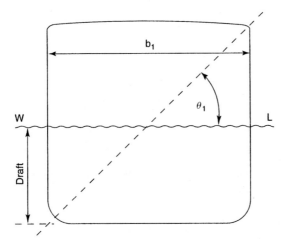

Figure 9.6

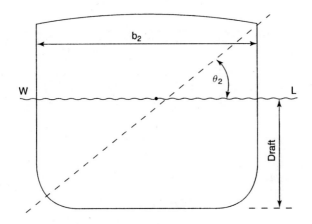

Figure 9.7

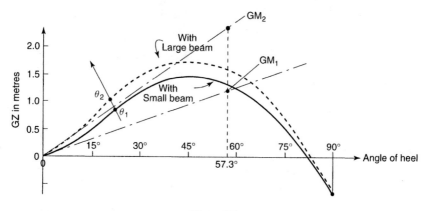

Figure 9.8

4. KB remains constant because draft does not change.

5. BM and KM increase due to extra width giving increased moment of Inertia at the water-plane

6. Range of Stability will increase but only slightly.

The case of a vessel having a negative GM in the upright position was discussed on page 64. The first part of her Statical Stability curve would appear as in Fig. 9.9.

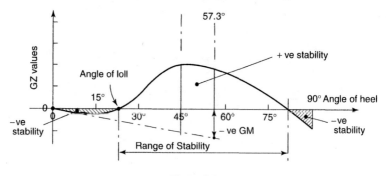

Figure 9.9

It is very important to observe that in Fig. 9.9, the Range of Stability is measured from the Angle of loll and *not* from the '0–0' axis.

Pages 70–72 refer to the term 'Angle of List'.

Pages 64 and 65 refer to the term 'Angle of Loll'.

Both angles are *fixed* angles of heel sometimes only for a few seconds or minutes.

What exactly is Angle of List and Angle of Loll? What are the characteristics of these two fixed angles of heel?

The following notes should help remove confusion of whether it is an Angle of List or Loll.

ANGLE OF LIST

'G', the centroid of the loaded weight, has *moved off the centre line* due to a shift of cargo or bilging effects, say to the port side.

GM is positive, i.e. 'G' is below 'M'. In fact GM will *increase* at the angle of list compared to GM when the ship is upright. The ship is in *stable equilibrium*.

In still water conditions the ship will remain at this *fixed* angle of heel. She will list to one side only, that is the same side as movement of weight.

In heavy weather conditions the ship will roll about this angle of list, say 3° P, but will not stop at 3° S. See comment below.

To bring the ship back to upright, load weight on the other side of the ship, for example if she lists 3° P add weight onto starboard side of ship.

ANGLE OF LOLL

KG = KM so *GM is zero*. 'G' remains *on the centre line* of the ship.

The ship is in *neutral equilibrium*. She is in a *more dangerous situation* than a ship with an angle of list, because once 'G' goes above 'M' she will capsize.

Angle of loll may be 3° P or 3° S depending upon external forces such as wind and waves acting on her structure. She may suddenly flop over from 3° P to 3° S and then back again to 3° P.

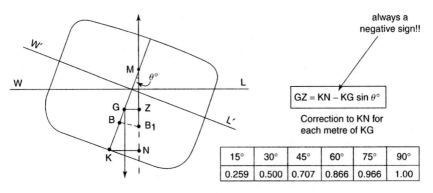

Figure 9.10

Ship Stability – Notes & Examples

To improve this condition 'G' must be brought below 'M'. This can be done by moving weight downwards towards the keel, adding water ballast in double-bottom tanks or removing weight above the ship's 'G'. Beware of free surface effects when moving, loading, and discharging liquids.

With an angle of list or an angle of loll the calculations must be carefully made *prior* to any changes in loading being made.

WORKED EXAMPLE 64

Draw a curve of statical stability, using the KN Cross Curves, for the vessel when displacing 9000 tonnes with a KG of 6.7 metres. Use Cross Curves in Fig. 9.11.

From the curve give the following.

(a) Range of stability.

(b) Change of the above range when a transverse upsetting moment of 2250 tonnes-metres is caused.

(c) Approximate GM.

(d) Moment of statical stability if heeled 5°.

(e) Approximate angle at which the deck edge immerses.

(f) Dynamical stability at 48°.

Erect a perpendicular line @ W = 9000 t. Lift off intersection value @ 15°, 30°, 45° etc.

Angle of heel	KN from Cross Curves	$-KG \sin\theta°$ i.e. $-6.7 \sin\theta°$	Actual GZ in metres for ship's condition.
0°	0	0	0
15°	1.98	1.73	0.25
30°	4.10	3.35	0.75
45°	5.92	4.74	1.18
60°	6.82	5.80	1.02
75°	6.98	6.47	0.51
90°	6.58	6.70	−0.12

Now replot corrected GZ values against Angle of Heel as shown in Fig. 9.12.

Correction to KN for each metre of KG for Cross Curves on next page.

θ	15°	30°	45°	60°	75°	90°
$\sin\theta$	0.259	0.500	0.707	0.856	0.966	1.000

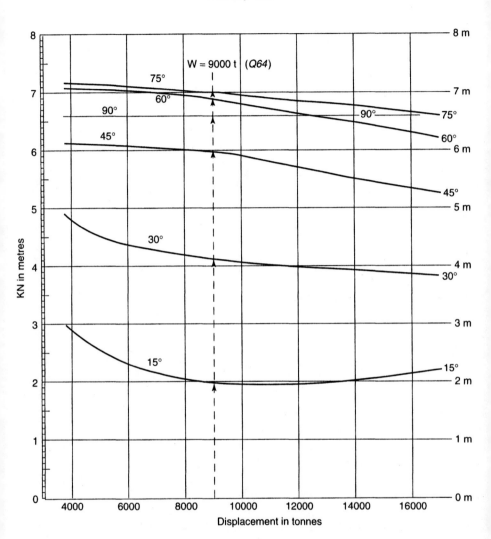

Figure 9.11 KN Cross Curves of stability

ANSWERS TO WORKED EXAMPLE 64

(a) 0°–86°. See Figure 9.12.

(b) Reduced to 14(1/2)°–80° (divide moment caused by the displacement to obtain reduction in GZ). See Fig. 9.12.

(c) 1.00 metre (extrapolated to $\theta = 57.3°$). See Figure 9.12.

(d) 785 tonnes-metres (i.e. W × GZ at 5°).

(e) 26° See Fig. 9.12.

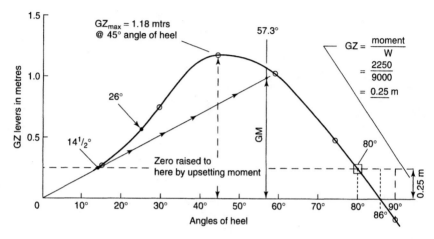

Figure 9.12

(f)

Heel	GZ in metres	SM	Area function
0°	0	1	0
8°	0.12	4	0.48
16°	0.30	2	0.60
24°	0.54	4	2.16
32°	0.82	2	1.64
40°	1.11	4	4.44
48°	1.17	1	1.17
			$10.49 = \sum_1$

$$\text{Dynamical stability} = \frac{1}{3} \times h \times \sum_1 \times W$$

$$= \frac{1}{3} \times 8°/573° \times 10.49 \times 9000$$

$$= 4394 \text{ tonnes-metres}$$

(Note: 48° can be divided into any convenient number of intervals, 6 or 8 making the calculation simple).

On page 58 it was stated that dynamical stability was the work done in inclining the vessel by external forces and that it was the product of the vessel's displacement and the vertical

Stability data

separation of B and G. Although this is perfectly true it is easier to calculate dynamical stability by the method shown above.

Think of the vessel heeled 1°, to heel her to 2° one would have to do some work to overcome the righting moment at 1°. If she has now been heeled to 2°, to heel her to 3° the righting moment at 2° must be overcome and the total moment to be overcome from the upright is the sum of the righting moments from 0 to 3°. This is true for any angle and so the dynamical stability at any angle is the area under a curve of Statical Stability up to that angle multiplied by the displacement.

It must be emphasised that the common interval must be expressed in circular measure and θ in circules measure is $\frac{\theta°}{57.3°}$. Also the tonnes-metres of dynamical stability are a measure of work done and are not the same units as the tonnes-metres by which moments are expressed.

Readers must be very careful when using cross curves of stability to note as to whether they are KN curves as shown on page 135 or GZ curves – the older method of presentation.

Summary

With GZ cross curves of Stability:

1. $GZ = GZ$ ordinate $\pm GG_1 . \sin\theta$; where $\theta = 0°$ to $90°$.

2. There is an ASSUMED KG given on the Curves eg ASSUMED KG = 8 m.

3. $GG_1 = KG_{ASSUMED} - KG_{ACTUAL}$.

4. GG_1 could be +ve, −ve or on rare occasions = zero.

With KN cross curves of Stability:

1. $GZ = KN$ ordinate $- KG_{ACTUAL} \times \sin\theta$; where $\theta = 0°$ to $90°$.

2. The cross curves are applicable to a KG of ZERO.

3. The correction will *always be negative*, thereby reducing risk of error in the calculation of final GZ values.

4. KN ordinates are much larger in comparison with GZ ordinates on a set of cross curves.

CHAPTER TEN

Carriage of Stability Information

The provision of stability information for the use of ship's personnel has been required for some years and the minimum requirements were formerly stated in notice M. 375. These requirements, with some additions, have now been incorporated in the 1968 Load Line Rules and are reproduced below.

1. The ship's name, official number, port of registry, gross and register tonnages, principal dimensions, displacement, deadweight and draft to the Summer load line (SLWL).

2. A profile view and, if the Board so require in a particular case, plan views of the ship drawn to scale showing with their names all compartments, tanks, storerooms and crew and passenger accommodation spaces, and also showing the amidships position.

3. The capacity and the centre of gravity (longitudinally and vertically) of every compartment available for the carriage of cargo, fuel, stores, feed water, domestic water or water ballast.

In the case of a vehicle ferry, the vertical centre of gravity of compartments for the carriage of vehicles shall be based on the estimated centre of gravity of the vehicles and not on the volumetric centres of the compartments.

4. The estimated total weight of (a) passengers and their effect and (b) crew and their effects, and the centre of gravity (longitudinally and vertically) of each such total weight. In assessing such centres of gravity passengers and crew shall be assumed to be distributed about the ship in the spaces they will normally occupy, including the highest decks to which either or both have access.

5. The estimated weight and the disposition and centre of gravity of the maximum amount of deck cargo which the ship may reasonably be expected to carry on an exposed deck. The estimated weight shall include in the case of deck cargo likely to absorb water the estimated weight of water likely to be so absorbed and allowed for in arrival conditions, such weight in the case of timber deck cargo being taken to be 15 per cent weight.

6. A diagram or scale showing the load line mark and load lines with particulars of the corresponding freeboards, and also showing the displacement, metric tons per centimetre immersion, and deadweight corresponding in each case to a range of mean drafts extending between the waterline representing the deepest load line and the waterline of the ship in light condition.

7. A diagram or tabular statement showing the hydrostatic particulars of the ship, including:

(1) the heights of the transverse metacentre and

(2) the values of the moment to change trim one centimetre,

for a range of mean drafts extending at least between the waterline representing the deepest load line and the waterline of the ship in light condition. Where a tabular statement is used, the intervals between such drafts shall be sufficiently close to permit accurate interpolation. In the case of ships having raked keels, the same datum for the heights of centres of buoyancy and metacentres shall be used as for the centres of gravity referred to in paragraphs 3, 4 and 5.

8. The effect on stability of free surface in each tank in the ship in which liquids may be carried, including an example to show how the metacentric height is to be corrected.

9. (1) A diagram showing cross curves of stability indicating the height of the assumed axis from which the Righting Levers are measured and the trim which has been assumed. In the case of ships having raked keels, where a datum other than the top of keel has been used the position of the assumed axis shall be clearly defined.

(2) Subject to the following sub-paragraph, only (a) enclosed superstructures and (b) efficient trunks as defined in paragraph 10 of Schedule 5 shall be taken into account in deriving such curves.

(3) The following structures may be taken into account in deriving such curves if the Board are satisfied that their location, integrity and means of closure will contribute to the ship's stability:

(a) superstructures located above the superstructure deck;

(b) deckhouses on or above the freeboard deck, whether wholly or in part only;

(c) hatchway structures on or above the freeboard deck.

Additionally, in the case of a ship carrying timber deck cargo, the volume of the timber deck cargo, or a part thereof, may with the Board's approval be taken into account in deriving a supplementary curve of stability appropriate to the ship when carrying such cargo.

(4) An example shall be given showing how to obtain a curve of Righting Levers (GZ) from the cross curves of stability.

(5) Where the buoyancy of a superstructure is to be taken into account in the calculation of stability information to be supplied in the case of a vehicle ferry or similar ship having bow doors, ship's side doors or stern doors, there shall be included in the stability information a specific statement that such doors must be secured weathertight before the ship proceeds to sea and that the cross curves of stability are based upon the assumption that such doors have been so secured.

10. (1) The diagram and statements referred to in sub-paragraph (2) of this paragraph shall be provided separately for each of the following conditions of the ship:-

(a) Light condition. If the ship has permanent ballast, such diagram and statements shall be provided for the ship in light condition both (i) with such ballast, and (ii) without such ballast.

(b) Ballast condition, both (i) on departure, and (ii) on arrival, it being assumed for the purpose of the latter in this and the following sub-paragraphs that oil fuel, fresh water, consumable stores and the like are reduced to 10 per cent of their capacity.

(c) Condition both (i) on departure, and (ii) on arrival, when loaded to the Summer loadline with cargo filling all spaces available for cargo, cargo for this purpose being taken to be homogeneous cargo except where this is clearly inappropriate, for example in the case of cargo spaces in a ship which are intended to be used exclusively for the carriage of vehicles or of containers.

(d) Service loaded conditions, both (i) on departure and (ii) on arrival.

(2) (a) A profile diagram of the ship drawn to a suitable small scale showing the disposition of all components of the deadweight.

(b) A statement showing the lightweight, the disposition and the total weights of all components of the deadweight, the displacement, the corresponding positions of the centre of gravity, the metacentre and also the metacentric height (GM).

(c) A diagram showing a curve of Righting Levers (GZ) derived from the cross curves of stability referred to in paragraph 9. Where credit is shown for the buoyancy of a timber deck cargo the curve of Righting Levers (GZ) must be drawn both with and without this credit.

(3) The metacentric height and the curve of Righting Levers (GZ) shall be corrected for liquid free surface.

(4) Where there is a significant amount of trim in any of the conditions referred to in sub-paragraph (1) the metacentric height and the curve of Righting Levers (GZ) may be required to be determined from the trimmed waterline.

(5) If in the opinion of the Board the stability characteristics in either or both of the conditions referred to in sub-paragraph (1)(c) are not satisfactory, such conditions shall be marked accordingly and an appropriate warning to the master shall be inserted.

11. Where special procedures such as partly filling or completely filling particular spaces designated for cargo, fuel, fresh water or other purposes are necessary to maintain adequate stability, a statement of instructions as to the appropriate procedure in each case.

12. A copy of the report on the inclining test and of the calculation therefrom of the light condition particulars.

The information on hydrostatic particulars mentioned in paragraph 7 could be presented in the form which is illustrated on page 143. In modern ships this data is presented in *tabular form* and as part of a computer package. See Worked Example 66.

The diagram, information and curve of statical stability reproduced on the opposite page show a typical way in which the information required by paragraph 10 above could be presented. This diagram will be seen to illustrate paragraph 10 (1) (d) (i) and a further series of diagrams would have to be supplied to illustrate the other conditions detailed in paragraph 10. The diagrams may be collated in booklet form or displayed in the form of a chart under glass on a bulkhead or a desk top. This booklet is 'The Trim & Stability Book' and is consulted frequently by ship-board officers on the Bridge.

A displacement scale in 'Notes on Cargo Work' shows how the information required in paragraph 6 could be presented.

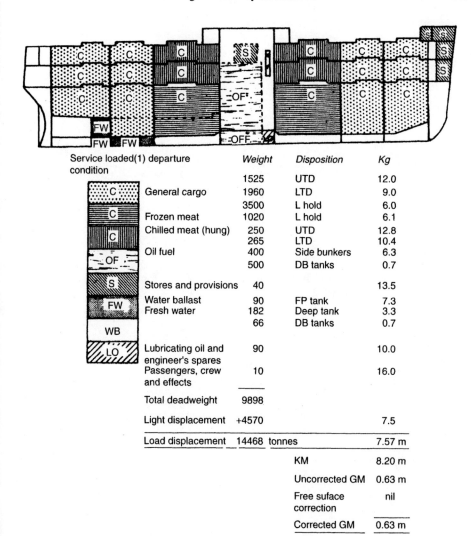

Service loaded(1) departure condition	Weight	Disposition	Kg
General cargo	1525	UTD	12.0
	1960	LTD	9.0
	3500	L hold	6.0
Frozen meat	1020	L hold	6.1
Chilled meat (hung)	250	UTD	12.8
	265	LTD	10.4
Oil fuel	400	Side bunkers	6.3
	500	DB tanks	0.7
Stores and provisions	40		13.5
Water ballast	90	FP tank	7.3
Fresh water	182	Deep tank	3.3
	66	DB tanks	0.7
Lubricating oil and engineer's spares	90		10.0
Passengers, crew and effects	10		16.0
Total deadweight	9898		
Light displacement	+4570		7.5
Load displacement	14468 tonnes		7.57 m
		KM	8.20 m
		Uncorrected GM	0.63 m
		Free suface correction	nil
		Corrected GM	0.63 m

Figure 10.1 Typical example of a page from a Trim and Stability Book supplied to ship by shipbuilders.

The Marine Division of the Department of Trade and Industry publish a Stability Information Booklet which indicates a recommended method of presenting the stability information to comply with The Merchant Shipping (Load Line) Rules 1968.

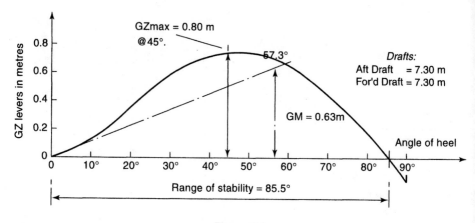

Figure 10.2

WORKED EXAMPLE 65

A vessel of length 156 m whose hydrostatic curves are shown in Fig. 10.4 is at present displacing 16 000 tonnes. Her longitudinal centre of gravity is known to be 3.0 metres abaft amidships.

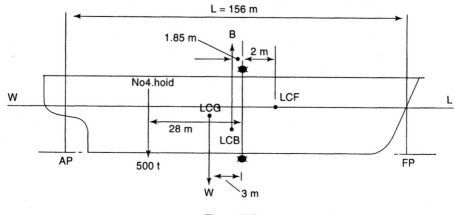

Figure 10.3

Estimate her draft after loading 500 tonnes of cargo into No. 4. hold whose centre of gravity is 28 m abaft amidships.

From the curves at a displacement of 16 000 tonnes the mean draft is 8.00 m. Then, as all information is plotted against draft, from Fig. 10.4 the following is obtained:

 LCF 2.00 m forward of amidships
 LCB 1.85 m abaft of amidships
 TPC 22.3
 MCT 1 cm 200 tonnes-metres

Carriage of stability information

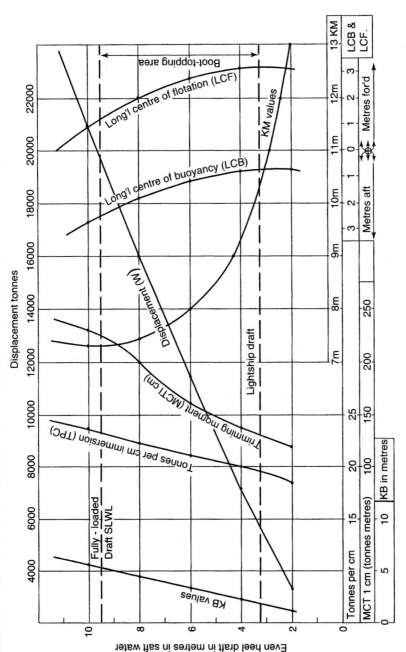

Figure 10.4 Hydrostatic Curves

Note: The above information would be for an even keel condition

If L.C.G. is 3.0 m abaft amidships
and L.C.B. is 1.85 m abaft amidships $\Big\}$ ∴ LCG to LCB = 1.15 m.

i.e. a distance of 1.15 m between them if the vessel is on an even keel which constitutes a lever. To bring the vessel into longitudinal equilibrium a moment of 16 000 × 1.15 m is required.

$$\text{Change of trim} = \frac{W \times (\text{LCG to LCB})}{\text{MCT 1 cm}}$$

This causes a trim of $\dfrac{16\,000 \times 1.15}{200}$ by the stern = 92 cm ↻

Trim due to loading 500 tonnes 30 m abaft LCF = $\dfrac{500 \times 30}{200}$ = 75 cm ↻

total trim 167 cm ↻ i.e. 1.67 m

Original mean draft	8.00 m	
Sinkage due to loading 500 tonnes = W/TPC = $\dfrac{500}{22.3}$ =	0.22 m	
New mean draft @ LCF position =	8.22 m	8.22 m
Trim ratio forward = $\dfrac{76 \times 1.67}{156}$ =	–	−0.81 m
Trim ratio aft = $\dfrac{80 \times 1.67}{156}$ =	+0.86 m	
Final end drafts	9.08 m A	7.41 m F

TABULATED HYDROSTATIC DATA

WORKED EXAMPLE 66

A ship 136 m LBP is loaded ready for departure. Her displacement is 12 756 tonnes with an LCG of 0.28 m aft of amidships. From her Trim and Stability Book it is known that:

Draft d_H	Displacement (t)	MCT 1 cm (t.m./cm)	LCF	LCB ⊗
8.00 m	14 253	162.9	1.20 m aft	⊗
7.00 m	12 258	158.1	0.50 m aft	0.45 m for'd

Carriage of stability information

Calculate, using the above hydrostatic information, her final departure drafts for'd and aft. Interpolate first within the displacement values, as follows:

$$\left.\begin{array}{l}14\,253 - 12\,258 = 1995\,\text{tonnes}\\ 12\,756 - 12\,258 = 498\,\text{tonnes}\end{array}\right\} \frac{498}{1995} = 0.25\,\text{m} = \delta d$$

$\therefore$ mean draft $= 7.00 + 0.25 = 7.25\,m$ @ LCF position.

At $7.25\,m$ draft; $MCTC = 158.1 + 0.25(162.9 - 158.1)$

$$= 159.3\,t \cdot m./cm.$$

$$LCF_{\cancel{x}} = 0.50 + 0.25(1.20 - 0.50) = 0.68\,\text{m aft} \cancel{x} \quad \text{i.e.} +0.68\,\text{m}.$$

$$LCB_{\cancel{x}} = 0.45 - (0.25 \times 0.45) = 0.34\,\text{m for'd} \cancel{x} \quad \text{i.e.} -0.34\,\text{m}.$$

$$\text{Change of trim} = \frac{W \times (LCG - LCB)}{MCTC} = \frac{12\,756 \times \overset{+0.62\,m}{(+0.28 - -0.34)}}{159.3} = +50\,\text{cms}$$

$$= 50\,\text{cms by the Stern}.$$

Draft aft = mean draft + trim ratio

$$= 7.25\,\text{m} + \left[\frac{67.32}{136} \times 0.50\,\text{m}\right]$$

$$= 7.50\,m.$$

$$Draft\,for'd = 7.25\,\text{m} - \left[\frac{68.68}{136} \times 0.50\,\text{m}\right]$$

$$= 7.00\,m$$

CONCLUDING REMARKS

In the past fifty years, mechanical and electrical instruments have been used onboard ships to evaluate many of the Stability and Strength values discussed in the ten chapters of this text book.

They include the following:

1. The 'Ralston' trim and stability evaluator.
2. The 'Kelvin Hughes' mechanical stress and trim indicator.
3. The 'Stabilologue', giving mean draft and GM with free surface effects included.
4. The 'Dobbie–McInnes Stabilograph'.
5. The electrical 'Loadicator', giving Shear Forces and Bending Moments along the ship's length.

Due to computer technology, these mechanical and electrical instruments are no longer used onboard ships. They have become obsolete. They have been replaced with software and hardware that can calculate the required values accurately and very quickly. Many other options can be also be considered in a short period of time.

As well as numerical output of data, computers can produce graphics showing SF and BM curves from bow to stern with maximum limit guidelines superimposed.

If an amount of cargo is being keyed in for a compartment and that compartment is not big enough to accept it, then an audio or flashing signal warns the programmer of the error. Similarly if a loaded condition contravenes D.Tp. regulations, for example a GM of less than 0.15 m then again a warning is given.

Further description of computer programmes and packages are outside the pre-requisites of this Ship Stability book.

APPENDIX I

Revision One-liners

The following are sixty five one-line questions acting as an aid to examination preparation. They are similar in effect to using mental arithmetic when preparing for a mathematics exam. Elements of questions may well appear in the written papers or in the oral exams... Good luck.

1. What is another name for the KG?
2. What is a Hydrometer used for?
3. If the angle of heel is less than 10 degrees, what is the equation for GZ?
4. What are the formulae for TPC and MCTC for a ship in salt water?
5. Give two formulae for the Metacentre, KM.
6. How may Free Surface Effects be reduced on a ship?
7. What is another name for KB?
8. List four requirements before an Inclining Experiment can take place.
9. With the aid of a sketch, define LOA and LBP.
10. What are Cross Curves of Stability used for?
11. What is the longitudinal centre of a waterplane called?
12. Adding a weight to a ship usually causes two changes. What are these changes?
13. What is Simpson's 1st rule for a parabolic shape with seven equally spaced ordinates?
14. What is KB for (a) box-shaped vessel and (b) triangular-shaped vessel?
15. What are Hydrostatic Curves used for onboard a ship?
16. Using sketches, define the Block, the Waterplane and Midship form coefficients.
17. Sketch a Statical Stability curve and label six important points on it.
18. What are the minimum values allowed by D.Tp., for GZ and for transverse GM?
19. List three ways in which a ship's end drafts may be changed.
20. GM is 0.45 m. Radius of gyration is 7 m. Estimate the natural rolling period in seconds.

21. What is a 'Datum' position or point?
22. What is the formula for Bending Stress in terms of M, I and y?
23. Sketch a set of Hydrostatic Curves.
24. List three characteristics of an Angle of Loll.
25. Define (a) a moment and (b) a moment of inertia.
26. Sketch the 1st three curves for a set of ship's Strength Curves.
27. What is the 'theory of parallel axis' formula?
28. Effects on a Statical Stability curve for increased Breadth, and increased Freeboard?
29. Sketch a Metacentric Diagram for a box-shaped vessel and a triangular-shaped vessel.
30. Block coefficient is 0.715 Midship coefficient is 0.988. Calculate Prismatic coefficient.
31. Describe the use of Simpson's 3rd rule.
32. What is the wall-sided formula for GZ?
33. Define 'permeability'. Give two examples relating to contents in a hold or tank.
34. Equations for BM, for box-shaped vessels and for triangular-shaped vessels.?
35. List three characteristics of an Angle of List.
36. Sketch the Shear force and Bending Moment curves Show their inter-relation.
37. For a curve of seven equally spaced ordinates give Simpson's 2nd rule.
38. What is the formula for pressure for water on a lockgate situation?
39. When a weight is lifted from a jetty by a ship's derrick whereabouts does its CG act?
40. What is a slack tank and how does it affect ship stability?
41. Sketch a Displacement curve.
42. What is Morrish's formula for VCB?
43. For an inclining experiment how is tangent of the angle of list obtained?
44. What do 'a moment of statical stability' and 'dynamical stability' mean?
45. Show the range of stability on a S/S curve having a very small initial negative GM.
46. Breadth is 45 m. What is the increase in draft at a list of 2 degrees?

47. What is the formula for loss of GM due to free surface effects in a slack tank?
48. For what purpose is the Inclining Experiment made on ships?
49. What is the 'true mean draft' on a ship?
50. When drydocking a ship there is a virtual loss in GM. Give two formulae for this loss.
51. With Simpson's rules, formulae for M of I about (a) Amidships (b) Centreline?
52. Discuss the components involved for estimating an angle of heel whilst turning a ship.
53. What is a 'stiff ship' and a 'tender ship'. Give typical GM values.
54. With the Lost Buoyancy method how does VCG change, after bilging has occurred?
55. What is a fulcrum?
56. Sketch a Bending stress diagram for a vessel that is in a Sagging condition.
57. Whereabouts on a ship is the 'boot-topping' area?
58. Define 'Ship Squat'.
59. What do the letters SLWL signify. Give another name for this term.
60. What is a Moment of Weights table?
61. What happens to cause a vessel to be in Unstable equilibrium?
62. What causes Hogging in a vessel?
63. Which letters signify the Metacentric Height?
64. Give typical Cb values for fully-loaded VLCC, General Cargo ships, Passenger Liners
65. What happens when a ship passes from one density of water to another water density?

APPENDIX II

Problems

1. A box-shaped vessel 100 m in length 15 m beam is floating at a draft of 5.3 m in water relative density 1.013. Calculate the number of tonnes she can load, if her maximum permitted draft in salt water is 6.0 m.

2. A vessel displacing 6800 tonnes is listed 5° to starboard with a GM of 0.7 m. Calculate how many tonnes should be loaded in the port tween deck 4 m to port of the centre line and 0.3 m above the old C.G. in order to bring the vessel upright.

3. A vessel of 12 500 tonnes KM 8.5 m KG 7.0 m is floating upright. She then loads the following: 400 t, KG 5 m; 770 t, KG 6.8 m; 300 t, KG 0.5 m; 835 t, KG 10 m. She discharges 460 t, KG 9 m; 565 t, KG 8 m. Calculate her list if 100 tonnes of oil is transferred from port to starboard, a distance of 14 metres.

4. A vessel which displaces 14 000 tonnes when floating at an even keel draft of 8.25 metres has a length on the waterplane of 144 metres. The half-ordinates of her waterplane at equidistant intervals are, starting from forward: 0, 2.7, 5.9, 8.3, 10.6, 10.6, 9.9, 6.4 and 2.1 metres respectively. Calculate the position of the centre of flotation, her fresh water allowance and her waterplane coefficient.

5. A double bottom tank 20 m long 16 m wide 1 m depth has a centreline longitudinal watertight division. Calculate the vessel's list if the port side of this tank is half filled with oil relative density 0.975. Assume the vessel's displacement in salt water, KM and KG before loading the oil to have been 9422 tonnes, 8.4 and 7.5 metres respectively.

6. A ship of length 150 m drawing 8.2 m forward 9.35 m aft with MCT 1 cm of 210 tonnes-metres and TPC 25, loads the following: 300 t, 40 m forward of C.F; 350 t, 52 m abaft C.F; 100 t at C.F; 225 t, 12 m forward of C.F. A further quantity of cargo is expected, space is available in the after hold 63 m abaft C.F. and in the forward tween decks. How much should be loaded aft if the draft aft is to be 10.0 m? Where should the remainder be stowed to maintain this after draft? The C.F. is 5 m abaft amidships.

7. A vessel drawing 6.65 m forward 7.33 m aft is required to complete loading at a mean draft of 8.00 m and be trimmed 0.5 m by the stern. Space is available 43 m forward of C.F. and 48 m abaft C.F. How much cargo should be loaded in each position to achieve the required draft? MCT 1 cm 175 tonnes-metres; TPC 18. C.F. is at amidships.

8. A vessel drawing 6.75 m forward and 7.95 m aft is to enter a channel with a maximum even keel draft of 7.4 m. The forepeak tank 60 m from the C.F. is the only empty tank and this may be filled, either from overside and/or by transferring water ballast from

Problems

No. 5 D.B. tank (capacity 120 tonnes) 40 m abaft C.F. How much should be run into and/or transferred to the forepeak to obtain the correct trim at maximum draft? MCT 1 cm 150 tonnes-metres; TPC 20.

9. A box-shaped vessel 125 metres in length 18 metres beam is floating at a draft of 9 metres with a KG of 7.0 m. A discharge pipe located at amidships on the starboard waterline is fractured. In order to make repairs it is to be brought 1 metre above the water level. How much oil should be shifted from side to side, a distance of 10 metres to achieve this? What were the waterplane area and the corresponding TPC?

10. A ship is floating on an even keel in salt water at a draft of 7.5 m. Her waterplane length, breadth and coefficient are 150 m, 16 m and 0.833 respectively. A double bottom tank 19 m long 15 m wide and 1.2 m deep is to be filled by opening the sea valve and allowing the water to run in. Calculate the maximum thrust on the tank top.

11. A vessel of 9300 tonnes, KM 9.25 metres is inclined by shifting 15 tonnes horizontally across the deck 21 metres. It is noted that the mean deflection of a plumbline 13.5 metres long is 35.5 centimetres. Calculate her KG and her angle of Inclination.

12. The tonnes per centimetre immersion of a vessel up to her load draft of 15.2 metres are given for equally spaced waterlines as follows:

$$0, \ 18.5, \ 19.4, \ 20.4, \ 21.6, \ 23.3, \ 24.7, \ 26.0, \ 27.5.$$

Calculate her displacement in salt water and the position of the centre of buoyancy, checking the latter by Morrish's Formula.

13. A vessel of length 186 m drawing 9.45 m forward and 10.35 m aft loads 200 t, 20 m forward of the C.F. and 375 t, 45 m abaft the C.F. She then discharges 95 t from a point 25 m abaft the C.F. Calculate the new end drafts if the C.F. is 5 m abaft amidships the moment to change trim 1 cm is 270 tonnes-metres and the TPC is 32.

14. A box-shaped vessel 150 metres in length and 20 metres beam is floating at a draft of 7 metres in salt water with a KG of 7.6 metres. A midships compartment extending across the vessel and 25 metres long is bilged. Calculate her moment of statical stability if heeled 5°.

15. A box-shaped vessel 150 metres long, 25 metres beam, is floating on an even keel in salt water at a draft of 6 metres. A midship compartment length 30 metres containing cargo stowing at 2.0 cubic metres per tonne, relative density 0.625, is bilged. Calculate the new draft.

16. A box-shaped vessel length 100 m, beam 12 m, depth 10 m is floating at a draft of 6.0 m in salt water. Her KG is 4.5 m. What is:

a) Her metacentric height?

b) The moment of statical stability when heeled 20°?

c) The dynamical stability when heeled 20°?

17. A vessel 216 metres in length has the following equally spaced ordinates on the bottom plating which is assumed to be flat. 0, 12, 16, 18.3, 19.6, 20, 18.1, 13.1, 6.9 and 1.9 metres. What is the upthrust on the bottom plating if the vessel is floating at a draft of 8 metres in water of relative density of 1.018?

18. A waterplane is defined by the following ordinates:

    0, 2.4, 3.3, 3.4, 2.9 and 0.6 metres. If the vessel is 15 metres long calculate the TPC at her present draft.

19. A vessel is floating at a draft of 7.3 m forward and 7.0 m aft. Given TPC 20 and MCT 1 cm 125 tonnes-metres, calculate how much cargo to load into No. 4. hold c.g. 45 m abaft the centre of flotation and No. 1. hold 60 m forward of the centre of flotation, to bring the vessel to an even keel at a draft of 7.5 m. Centre of flotation is at amidships.

20. A vessel of light displacement 3500 tonnes KG 6.5 m KM 7.2 m has to load 9000 tonnes of ore. KG of lower hold 6.0 m and tween deck 13.0 m. If the only requirement is to have a righting moment of 500 tonnes-metres at 8° when loaded, how much cargo should be loaded into each available position?

21. A ship 200 m in length, displacement 12 200 tonnes leaves port on an even keel. She consumes 600 tonnes of fuel KG 0.75 m; 8 m forward of the C.F. and 150 tonnes of water KG 6.0 m; 96 m forward of C.F. Calculate the quantity of water to transfer from the after peak (cap. 100 t) to the fore peak a distance of 170 m and what to load, if necessary, into a D.B. tank 40 m forward of C.F. KG 0.8 m to bring the vessel back to even keel.

22. A ship with drafts F 10.5 m A 13.0 m has to berth where there is only 13.0 m of water alongside. Working on a minimum clearance of 0.2 m under the keel, calculate the ballast to take 25 m forward of the C.F. which is 2 m abaft amidships. Length 180 m, TPC 18, MCT 1 cm 100.

23. A ship displacing 9980 tonnes GM 0.8 m KG 11.0 m lifts a container weighing 20 tonnes. Find the list when the container is first lifted off the jetty. The derrick is plumbed 16 m outboard and the head block is 16 m above the jetty and 21 m above the keel. Find also the list when the container is placed aboard having a KG 9 m and 9 m outboard of C/L.

24. A beam 20 m long of negligible weight is loaded with 20 tonnes evenly spread. Draw a shear force and bending moment curve and find the values of both 6 m from one end, if the beam is supported on knife edges at each end.

25. A cantilever 10 m long of negligible weight has a 5 tonne weight attached at the free end. Draw S.F. and B.M. curves and find shear and the bending moment at amidships.

26. A vessel of length 120 metres, draft 6.2 m forward and 7.0 m aft has MCT 1 cm 100 and TPC 12. Find the weight of water to pump out of the after peak so that she can pass over a bar with depth of water 6.9 m with a clearance of 20 centimetres. The after peak is 50 m aft of the CF which is 2 m abaft amidships.

Problems

27. The 1/2 ordinates of a vessel's waterplane are 0; 2.4; 5.4; 7.2; 7.8; 9.0; 9.6; 8.4; 7.2; 4.2; 0. If the waterplane is 140 m long find the TPC at this draft in fresh water and the position of the centre of flotation. Also calculate C_w value.

28. A box-shaped vessel length 20 m breadth 9 m depth 7 m floating in freshwater on an even keel of 2.0 m has a KG 4.0 m, and loads 540 tonnes of concentrate spread evenly with a KG of 3.0 m. Calculate the original GM and also the loaded GM.

29. A ship displacing 10 000 tonnes has a GM 1.0 m and is listed 4° to starboard. It is required to load a further 250 tonnes KG 10.0 m. Assume KM of 12.0 m is constant. Space is available 6.0 m to starboard of centre line and 4.0 m to port of centre line. How much cargo should be loaded into each if the vessel is to be upright on completion?

30. Estimate the amount of cargo aboard a vessel 150 m in length, draft of 6.6 m F 9.8 m A, centre of flotation 2 m abaft amidships and stores and fuel 300 tonnes. The displacement curve gave the following information:

Draft in metres	Displacement in tonnes
8.4 (loaded)	11 090
8.0	10 570
2.5 (light)	3 700

31. A transverse bulkhead has water ballast on one side to a depth of 12 m and oil on the other side to a depth of 18 m. The bulkhead is rectangular and 16 m wide. Density of the water ballast is 1.025 t/cu.m. and density of the oil is 0.895 t/t/cu.m.

    Calculate the resultant thrust and its centre of pressure above base.

32. A vessel's waterplane has the following half-ordinates spaced 20 m apart commencing at the After Perpendicular:

    0.50, 6.00, 12.00, 16.00, 15.00, 9.00 and 0 metres.

    For this waterplane, calculate the following information:

    (a) Waterplane area P&S.
    (b) TPC in river water, of density 1.012 t/cu.m.
    (c) LCF from amidships.
    (d) Moment of Inertia about amidships.
    (e) Moment of Inertia about LCF.
    (f) Moment of Inertia about the Centreline.

33. The draft marks of a vessel 120 m LBP show that the Aft draft reading is 5.15 m whilst the Forward draft reading is 4.05 m. If the Aft draft marks are 5 m for'd of the AP and the for'd draft marks are 4.00 m aft of the FP, then calculate the corresponding drafts at the AP and the FP.

154 Ship Stability – Notes & Examples

34. The following data was lifted from a set of KN Cross Curves of Stability

Angle of Heel	0°	15°	30°	45°	60°	75°	90°
KN ordinate (m)	0	3.20	6.50	8.75	9.70	9.40	8.40

If the ship's actual KG was 9.00 m then calculate the GZ ordinates and plot the Statical Stability curve. From it evaluate;

(a) Range of Stability. (b) Angle of Heel at which Deck immersion takes place
(c) Maximum GZ. (d) Angle of Heel at which maximum GZ occurs.
(e) Righting Moment when Angle of Heel is 20°.

35. A ship has a displacement of 4650 tonnes, TPC of 14, a KM (considered fixed) of 4.5 m and a KG of 3.3 m. She is on even keel. She enters dry-dock and is set on the blocks.

(a) Calculate her effective GM (using two methods) when a further 0.50 m of water is pumped from the dock.

(b) Prove (using two methods) that the Righting Moment is 2433 tonnes.mtrs.

36. For a ship the Breadth Moulded is 21.75 m, KG is 6.82 m and KM is 8.25 m.

(a) Using an approximation for the radius of gyration proceed to calculate the natural rolling period T_R for this condition of loading.

(b) Estimate T_R, using an approximate formula.

(c) Based on the first two answers, discuss if this is 'a stiff ship' or a 'a tender ship'.

37. A box-shaped vessel is 80 m long, 12 m wide and floats at 3 m even keel in fresh water. It is equally divided into five compartments by transverse bulkheads. 150 t of cargo are already in No. 2 Hold, 50 t in No. 4 Hold and 50 t in No. 5 Hold.

(a) Draw the Shear Force and Bending Moment diagrams.

(b) Determine the maximum SF and BM values, showing whereabouts they occur.

38. A box shaped vessel has the following data.

Length is 80 m, breadth is 12 m, draft even keel is 6 m, KG is 4.62 m.

A double bottom tank 10 m long, of full width and 2.4 m depth is then half-filled with water ballast having a density of 1.025 t/m^3. The tank is located at amidships.

Calculate the new even keel draft and the new transverse GM after this water ballast has been put in the double bottom tank.

39. A box shaped vessel is 60 m long, 13.73 m wide and floats at 8 m even keel draft in salt water.

(a) Calculate the KB, BM and KM values for drafts 3 m to 8 m at intervals of 1 m. From your results draw the Metacentric Diagram.

(b) At 3.65 m draft even keel, it is known that the VCG is 4.35 m above base. Using your diagram, estimate the transverse GM for this condition of loading.

(c) At 5.60 m draft even keel, the VCG is also 5.60 m above base. Using your diagram, estimate the GM for this condition of loading. What state of equilibrium is the ship in?

40. A ship is 130 m LBP and is loaded ready for Departure as shown in the table below. From her Hydrostatic Curves at 8 m even keel draft in salt water it is found that:

MCTC is 150 t.m./cm, LCF is 2.5 m for'd ⋈, W is 12 795 tonnes and LCB is 2 m for'd ⋈.

Calculate the final end drafts for this vessel. What is the final value for the trim? What is the final Dwt value for this loaded condition?

Item	Weight in tonnes	LCG from midships
Lightweight	3600	2.0 m aft
Cargo	8200	4.2 m for'd
Oil Fuel	780	7.1 m aft
Stores	20	13.3 m for'd
Fresh water	100	20.0 m aft
Feed water	85	12.0 m aft
Crew and effects	10	at amidships

41. A ship has a displacement of 9100 tonnes, LBP of 120 m, even keel draft of 7 m in fresh water of density of $1.000 \, t/m^3$.

From her Hydrostatic Curves it was found that

$MCTC_{SW}$ is 130 t.m./cm, TPC_{SW} is 17.3 t, LCB is 2 m for'd ⋈ and LCF is 1.0 aft ⋈.

Calculate the new end drafts when this vessel moves into water having a density of $1.020 \, t/m^3$ without any change in the ship's displacement of 9100 tonnes.

42. A ship is just about to lift a weight from a jetty and place it on board. Using the data given below, calculate the angle of heel after the weight has just been lifted from this jetty. Weight to be lifted is 140 t with an outreach of 9.14 m.

Displacement of ship prior to the lift is 10 060 tonnes.

Prior to lift-off, the KB is 3.4 m, KG is 3.66 m, TPC_{SW} is 20, I_{NA} is 22 788 m^4, draft is 6.7 m in salt water. Height to derrick head is 18.29 m above the keel.

43. A ship of 8,000 tonnes displacement is inclined by moving 4 tonnes transversely through a distance of 19 m. The average deflections of two pendulums, each 6 m long was 12 cm. 'Weights on' to complete this ship were 75 t centred at Kg of 7.65 m. 'Weights off' amounted to 25 t centred at Kg of 8.16 m.

   (a) Calculate the GM and angle of heel relating to this information, for the ship as inclined.

   (b) From Hydrostatic Curves for this ship as inclined, the KM was 9 m. Calculate the ship's final Lightweight and VCG at this weight.

44. (a) Write a brief description on the characteristics associated with an 'Angle of Loll'.

   (b) For a box-shaped barge, the breadth is 6.4 m, draft is 2.44 m even keel, with a KG of 2.67 m.

   Using the given wall-sided formula, calculate the GZ ordinates up to an angle of heel up of 20°, in 4° increments. From the results construct a Statical Stability curve up to 20° angle of heel. Label the important points on this constructed curve.

   $$GZ = \sin\theta(GM + 1/2.BM.\tan^2\theta)$$

45. A beam 10 m in length is simply supported at its ends at R_A and R_B. Placed upon it is the following loading:

   3 tonnes at 3 m from R_A.

   2 tonnes at 5 m from R_A.

   4 tonnes at 7 m from R_A.

   1 tonne at 8 m from R_A.

   A spread load along the full 10 m, at 2 tonnes/m run.

   (a) Calculate the Reaction values at R_A and R_B.

   (b) Draw the Shear Force and Bending Moment diagrams.

   (c) Calculate the maximum Bending Moment.

46. A vessel is loaded up ready for Departure. KM is 11.9 m. KG is 9.52 m with a displacement of 20 550 tonnes. From the ship's Cross Curves of Stability, the GZ ordinates for a displacement of 20 550 tonnes and a VCG of 8 m above base are as follows:

Angle of heel. (θ)	0	15	30	45	60	75	90
GZ ordinate. (m)	0	1.10	2.22	2.60	2.21	1.25	0.36

Using this information, construct the ship's Statical Stability curve for this condition of loading and determine the following

(a) Maximum righting lever GZ.

(b) Angle of heel at which this maximum GZ occurs.

(c) Angle of heel at which the deck edge just becomes immersed.

(d) Range of Stability.

47. A ship has an LBP of 130 m, a displacement of 9878 tonnes with an LCG 0.36 m for'd of amidships. From her tabulated hydrostatic data it is known that:

Draft in metres	Displacement in tonnes	MCT 1 cm t.m./cm.	LCF from amidships	LCB from amidships
6.0	10 293	152.5	0.50 m for'd	0.80 m for'd
7.0	8 631	145.9	0.42 m for'd	1.05 m for'd

Calculate the end drafts For'd and Aft for this condition of loading.

48. A ship with a transverse metacentric height of 0.40 m has a speed of 21 knots, a KG of 6.2 m, with a centre of lateral resistance (KB) of 4.0 m.

The rudder is put hard over to Starboard and the vessel turns in a circle of 550 m radius. Considering only the centrifugal forces involved, calculate the angle of heel produced as this ship turns at the given speed.

49. A vessel has the following particulars

Displacement is 9,000 tonnes, natural rolling period is T_R of 15 seconds, GM is 1.20 m.

Determine the new natural rolling period after the following changes in loading have taken place

2,000 tonnes added at 4.0 m above ship's VCG.

500 tonnes discharged at 3.0 m below ship's VCG.

Assume that KM remains at same value before and after changes of loading have been completed. Discuss if this final condition results in a 'stiff ship' or a 'tender ship'.

50. Part of an Upper Deck has three half-ordinates spaced 15 m apart. They are 7.35, 9.76, and 10.29 m respectively. Calculate the area enclosed P&S and its LCG from first ordinate for the portion between the first two given offsets.

APPENDIX III

Answers to Problems

1. 1171.65 tonnes (1187 t if FWA is used)
2. 104.125 tonnes
3. 3°33′
4. 80.41 m from forward. FWA 170.5 mm Cw 0.656
5. 2°09½′ to port
6. 101.2 tonnes; 18 m forward of C.F.
7. 993.5 t forward; 824.5 t aft
8. Transfer 120 tonnes. Load 100 tonnes
9. 115.3 tonnes
10. 1890.34 tonnes, 1999 m², 20.49 t
11. 7.962 metres, 1.51°
12. 32 407.7 tonnes; 8.464 metres; 8.739 metres
13. 9.395 m forward: 10.684 m aft
14. 1065.5 tonnes-metres
15. 6.25 metres
16. (a) 0.5 m (b) 1596.4 tonnes-metres

    (c) 250 tonnes-metres approximately
17. 24 737.4 tonnes
18. 0.395
19. 264.3 t in No. 1 and 435.7 t in No. 4
20. 1379.62 t in T.D. and 7620.38 t in L.H.

Answers to problems 159

21. Transfer all A.P. and load 55 tonnes
22. 300 tonnes
23. 2°21' : 1°17'
24. S.F. 4 tonnes +ve. B.M. 42 t-m +ve
25. S.F. 5 tonnes +ve. B.M. 25 t-m −ve
26. 92.3 tonnes
27. 17.248; 74 metres from 1st ordinate, 0.642
28. 0.375 m; 0.45 m
29. 219.93 tonnes to port; 30.07 tonnes to starboard
30. 6885.5 tonnes
31. Resultant Thrust is 1139.04 tonnes, Centre of Pressure 8.07 m above base
32. 2380 sq.mtrs, 24.09 tonnes, 3.19 m for'd amidships, 1 592 000 m^4, 1 567 781 m^4, 134 978 m^4.
33. 5.20 m aft, 4.01 m for'd
34. 83.75°, approximately 15°, 2.39 m, 45°, 46,200 t.m
35. upthrust P is 700 t, 0.523 m or 0.616 m
36. 12.77 secs, 12.73 secs, a 'stiff ship' because rolling period <18 to 22 secs
37. SF = 50 t at 16 m & 32 m from the Stern, BM = 600 t.m. at 24 m from the Stern
38. 6.15 m, +0.27 m
39. GM = +1.8 m so vessel is stable, GM is zero so vessel is in neutral equilibrium
40. Aft draft is 8.23 m, for'd draft is 7.79 m, dwt is 9,195 t, 0.44 m Trim by the stern
41. Aft draft is 6.91 m, for'd draft is 6.87 m
42. 3.8°
43. 0.48 m, 1.13°, 8050 t, 8.51 m
44. Description, drawing and labelling solution
45. Maximum BM is 41.5 t.m. occurring at mid-length
46. 1.60 m, 40°, 12°, 72.5°

47. Aft draft 5.92 m, For'd draft 5.59 m
48. 6.8° to Port
49. T_R is 28.2 secs... rather a 'tender ship' because rolling period > 18 to 22 secs
50. 261.2 sq.mtrs, lcg is 7.85 m for'd of the first half-ordinate

APPENDIX IV

How to pass exams in Maritime Studies

To pass exams you have to be like a successful football team. You will need:

Ability... tenacity... consistency... good preparation... and luck!! The following tips should help you to obtain extra marks that could turn that 36% into a 42%+ pass or an 81% into an Honours 85%+ award. Good luck.

IN YOUR EXAM

1. Use big sketches. Small sketches tend to irritate Examiners.
2. Use coloured pencils. Drawings look better with a bit of colour.
3. Use a 150 mm rule to make better sketches and a more professional drawing.
4. Have big writing to make it easier to read. Make it neat. Use a pen rather than a biro. Reading a piece of work written in biro is harder to read especially if the quality of the biro is not very good.
5. Use plenty of paragraphs. It makes it easier to read.
6. Write down any data you wish to remember. To write it makes it easier and longer to retain in your memory.
7. Be careful in your answers that you do not suggest things or situations that would endanger the ship or the onboard personnel.
8. Reread you answers near the end of the exam. Omitting the word NOT does make such a difference.
9. Reread your question as you finish each answer. Don't miss for example part (c) of an answer and throw away marks you could have obtained.
10. Treat the exam as an advertisement of your ability rather than an obstacle to be overcome. If you think you will fail, then you probably will fail.

BEFORE YOUR EXAM

1. Select 'bankers' for each subject. Certain topics come up very often. Certain topics you will have fully understood. Bank on these appearing on the exam paper.

2. Don't swot 100% of your course notes. Omit about 10% and concentrate on the 90%. In that 10% will be some topics you will never be able to be understand fully.

3. Work through passed exam papers in order to gauge the standard and the time factor to complete the required solution. Complete and hand in every set Coursework assignment.

4. Write all formulae discussed in each subject on pages at the rear of your notes.

5. In your notes circle each formula in a red outline or use a highlight pen. In this way they will stand out from the rest of your notes. Remember formulae are like spanners. Some you will use more than others but all can be used to solve a problem.

6. Underline in red important key phrases or words. Examiners will be looking for these in your answers. Oblige them and obtain the marks.

7. Revise each subject in carefully planned sequence so as not to be rusty on a set of notes that you have not read for some time whilst you have been sitting other exams.

8. Be aggressive in your mental approach to do your best. If you have prepared well there will be a less nervous approach and like the football team you will gain your goal.

Index

After peak tank 100–101
Angle of Heel 85
  whilst turning 87–88
Angle of List 133 *see also* listing
Angle of Loll 63, 64, 65–66, 133–134
Appendage, stern 5
Archimedes' Principle 2
Area, centre of 24, 26–28
Area calculations 19–23
Atwood's formula 59

BM (bending moments) xi, xii, 34–35, 36, 37–46, 48–49, 51–54, 64, 130–132
Bilge radius 7
Bilging 13, 14, 15, 103
Block coefficient 4, 7–9, 11, 12
Boot-topping area 143
Bulbous bows 49–50
Buoyancy 46
  centre of 3, 36, 51, 103, 104
  curve of 45–46
  lost 13, 14, 103–104
  reserve 3, 9, 13

Cantilever 43–44
Carriage of stability information 138–146
Centres
  of area 24, 26–28
  of buoyancy 3, 36, 51, 103, 104 *see also* LCB
  of flotation 25–28, 34–35, 89, 95–98, 101 *see also* LCF
  of gravity 3, 16–18, 24, 65, 68–73, 76, 117, 138, 140 *see also* LCG
  of pressure xiii, 112, 113–114, 115–118
Centrifugal force 87, 88
Change of trim xiii, 89–94, 99–101, 103
Coefficient
  block 4, 7–9, 11, 12
  deadweight 6, 13, 31–32
  midship area 5, 7
  prismatic 5–6, 7–8, 11, 12–13
  waterplane area 4–5, 7–8, 11, 12, 24–25
Coefficients, design 4–6

Common intervals 19, 34
  sub-divided 32–34
Compartments, sub-divided 121–123
Computer technology 145–146
Conditions for stable equilibrium 62–63
Container ships in high wind conditions 130
Correction for layer 101
Correction to mean draft 101–102
Couple 57
Cross curves of stability 134–137, 139
Curves
  of buoyancy 45–46
  of loads 45–46
  of metacentres 53
  of statical stability *see* statical stability curves
  of upthrusts 45
  of weights 45
Curves, ship strength 49

DWT (Deadweight) xv, 31–32, 140
Datum point 18
Deadweight (DWT) xv, 31–32, 140
Deadweight coefficient 6, 13, 31–32
Deadweight-Moment curve 79–81
Deck cargo, weight of 138
Deck cargoes, timber 65, 138, 139
Density 2, 77
Density, relative 2–3
Derricks 74
Design coefficients 4–6
Discharging weights 66–67, 69, 71, 73, 98–99
Displacement 7–8, 9, 11, 29–30, 31–32, 140
Doors, ships with 139
Draft 84, 89, 92–93, 94–95, 102–103, 144–145
  change of density 104–106
  correction to mean 101–102
  when heeled 84–85
Dry-docking xiii, 107–110
Dynamical stability 58, 126, 127–129, 134–137

Effect of Free Surface (FSE) 119–124, 139
Equilibrium 62–64
Exams, how to pass 161–162

Index

FSE (Free Surface Effect) 119–124, 139
FWA (Fresh Water Allowance) 10, 12
Fishing vessels, icing up of 65
Flooding (bilging) 13, 14, 15, 103
Force, SI units of 1
Fore peak tank 50, 99–100, 116–117
Formulae for ship stability xi–xiii
Free Surface Effect (FSE) 119–124, 139
Free Surface Movement 124
Freeboard 129, 130
Fresh Water Allowance (FWA) 10, 12

GM (Greater Moment) 99, 126, 127–129, 134–136
  loss of 107, 109–110
  Loss of, due to FSE (Free Surface Effect) xiii, 121–124
  Typical values 54–57, 82
GZ cross curves of stability 137
GZ lever 57, 58, 60, 110, 120, 122–123, 125–130, 140

Half-ordinates *see* semi-ordinates
Hogging condition 3, 45
Hydrostatic curves 142–144
Hydrostatic data, tabulated 144–145

Inclining Experiment 76–78
Inertia, Moments of 35, 48, 51
Initial stability 58
Instruments 145

KM values xvii, 54–58
KN cross curves xiii, 134–137

LCB (longitudinal centre of buoyancy) 29
LCF (longitudinal centre of flotation) 25–28, 34–35, 89, 95–98, 101
LCG (longitudinal centre of gravity) 26–27, 32–34
Length, SI units of 1
Lever 57 *see also* Righting lever
Lightweight 76, 140
Listing 70–73, 74–76 *see also* Angle of List
Load Line Rules, 1968: 123, 124, 138–140
Loading weights 46, 66–68, 69–71, 72–73, 96–97, 102–103, 142–144
Loads, curve of 45–46

Longitudinal centres
  of buoyancy (LCB) 29
  of flotation (LCF) 25–28, 34–35, 89, 95–98, 101
  of gravity (LCG) 26–27, 32–34
Longitudinal metacentre 51
Loss of GM (Greater Moment) 107, 109–110
Loss of GM due to FSE (Free Surface Effect) xiii, 121–124
Loss of ukc (under keel clearance) xii, 84, 85–86
Lost buoyancy 13, 14, 103–104

MCT 1 cm (MCTC) 12–13, 90–91
Maritime exams, how to pass 161–162
Mass, SI units of 1
Metacentre 51, 58, 64, 140
Metacentre, transverse 51
Metacentres, curve of 53
Metacentric diagrams 54–57
Metacentric height 51, 53, 58, 126, 140
Metacentric stability 58
Midship area 11, 12
Midship area coefficient 5, 7
Midship compartment, bilged 13, 14, 15
Moment 1–2, 16, 34–35, 74–76 *see also* bending moments (BM)
  of statical stability 57, 59, 67, 134–136 *see also* righting moments
  of weights 67
Moments, principles of taking 16–18
Moments, righting xiii, 57, 61, 81, 110, 130
Moments about the centre of gravity 68–73
Moments about the keel 66–68
Moments of Inertia 35, 48, 51
Morrish's formula xi, 3, 31
Moseley's formula 59, 60

Negative stability 125, 126, 132
Neutral equilibrium 63

Offsets *see* semi-ordinates
Oil pressure 115–116

Parallel axis theorem 35
Passenger ships in high wind conditions 130
Passenger weight 138
Period of roll 81–84
Permeability 14, 15–16
Pressure 2, 111 *see also* upthrust
Pressure, centre of xiii, 112, 113–114, 115–118

Index

Pressure head 111, 112
Prismatic coefficient 5–6, 7–8, 11, 12–13

Radius of gyration 82
Range of stability 126, 127–129, 132, 134–136
Relative density 2–3
Relative positions of B, G and M 62
Reserve buoyancy 3, 9, 13
Resultant thrust 113–114, 115–116
Revision one-liners 147–149
Revision problems 150–157
   answers 158–160
Righting lever (GZ) 57, 58, 60, 110, 120, 122–123, 125–130, 140
Righting moments xiii, 57, 61, 81, 110, 130
Rise of floor 85
Rolling period 81–84

SI units 1–2
Sagging condition 3, 37, 45, 49
Semi-ordinates 19, 34
Shear forces 36–49
Shearing 36
Shift of B 52, 103
Shift of G 69, 70, 71, 74
Shifting weights 69, 71
Ship Squat – maximum values 86–87
Ship stability, concept of xvii
Ship strength curves 49
Ship strength diagrams ~ weight buoyancy and load 47
Ship surgery 7–8
Ship types and characteristics xv–xvi
   container ships in high wind conditions 130
   fishing vessels, icing up of 65
   passenger ships in high wind conditions 130
   ships with doors 139
   vehicle ferries, centre of gravity of 138
Simpson's Rules 19–20, 35
   areas 20–25
   centres xi, 25–26
   First Rule 20, 23, 25–26
   moments of inertia xi
   Second Rule 20–21, 24–25
   Third Rule 21, 26
   volumes xi, 22
Sinkage 13, 14–15, 105
Slack tanks 119
Specific gravity *see* relative density
Stabilisers 83
Stability
   cross curves of 134–137, 139
   dynamical 58, 126, 127–129, 134–137
   initial 58

   metacentric 58
   negative 125, 126, 132
   range of 126, 127–129, 132, 134–136
   statical 57
   wall-sided 60–61
Stability at large angles 58–60
Stability at small angles 51–58
Stability information, carriage of 138–146
Stability Information Booklet 141
Stability Test 76–78
Stable equilibrium 62–63
Statical stability 57
Statical stability curves 125–130, 132, 134–136, 140, 142
   effect of beam on 130–132
   effect of freeboard on 129–130
Stern appendage 5
Sub-divided common intervals 32–34
Sub-divided compartments 121–123
Suspended weights 74
Synchronism 81, 82–83

TPC (tonnes per centimetre immersion) 10, 12, 22, 24–25
Thrust, resultant 113–114, 115–116
Thrust due to liquid 112, 113–118
Timber deck cargoes 65, 138, 139
Tipping centre 89 *see also* LCF
Transverse metacentre 51
Trapezoidal Rule 19
Trim 89–91, 104
Trim, change of xiii, 89–94, 99–101, 103
Trim and Stability Book 140, 141
Trim ratio for'd & aft 89–91, 93–94, 95, 97–98
True mean draft 101–102

Under keel clearance xii, 84, 85–86
Unstable equilibrium 63, 64
Unsymmetrical loading 70–73
Upthrust 36, 114–115 *see also* pressure
Upthrust when dry-docking 107–108
Upthrusts, curve of 45

VCB (vertical centre of buoyancy) 29–31
Vehicle ferries, centre of gravity of 138
Vertical centre of buoyancy (VCB) 29–31

WPA (waterplane area) 11–12, 19, 22, 24–25, 34–35
Wall-sided formula 59, 60, 61, 65
Wall-sided stability 60–61

Waterplane area (WPA) 11–12, 19, 22, 24–25, 34–35
Waterplane area coefficient 4–5, 7–8, 11, 12, 24–25
Watertight flat 13, 14
Wave period 81, 82
Weight, SI units of 1
Weight loaded off centre line 70–73

Weight of deck cargo 138
Weights
  curve of 45
  discharging 66–67, 69, 71, 73, 98–99
  loading 46, 66–68, 69–71, 72–73, 96–97, 102–103, 142–144
  moment of 67
  suspended 74

Printed in the United Kingdom
by Lightning Source UK Ltd.
109360UKS00001B/61-78